listening to trees

A.K. Hellum

listening to trees

NeWest Press

Library and Archives Canada Cataloguing in Publication
Hellum, A.K., 1933–
Listening to trees / A.K. Hellum.
Includes bibliographical references and index.
ISBN 978-1-897126-33-2

1. Hellum, A. K., 1933– 2. Deforestation. 3. Forest protection. 4. Foresters--Canada--Biography. 5. Environmentalists--Canada--Biography. I. Title.
SD129.H44A3 2008   634.9092   C2008-902311-0

Editor for the Board: Don Kerr
Cover and interior design: Natalie Olsen
Cover illustration: Natalie Olsen
Interior illustrations: A.K. Hellum
Proof and index: Carol Berger

NeWest Press acknowledges the support of the Canada Council for the Arts, the Alberta Foundation for the Arts, and the Edmonton Arts Council for our publishing program. We also acknowledge the financial support of the Government of Canada through the Book Publishing Industry Development Program (BPIDP).

NeWest Press
201.8540.109 Street
Edmonton, Alberta T6G 1E6
780.432.9427
newestpress.com

No bison were harmed in the making of this book.
We are committed to protecting the environment and to the responsible use of natural resources. In our attempt to listen to trees, this book is printed on 100% recycled, ancient forest-friendly paper.

1 2 3 4 5 11 10 09 08
printed and bound in Canada

table of Contents

"Pinus strobus"
"Eastern White Pine"
Menomonie, WI.
June 16, 2006

Preface

Listening to Trees is a collection of ideas and stories gathered throughout a professional lifetime. They tell of my experiences working with people, forests, and trees.

I came from the city, found and fell in love with work in the forests of Norway in the 1950s, studied forestry at universities in North America, and later taught forestry in Alberta. Before and after retirement I consulted abroad and at home.

The longer I worked in forestry, however, the more disillusioned I became with my chosen profession. I became increasingly disturbed by what I perceived as a callous and uncaring use of forest almost everywhere I went after leaving Norway. Now the practice of forestry in Norway is much the same as everywhere; people use the same huge machines and leave cutovers in disturbed and often very messy conditions, almost everywhere you care to look.

Because forestry practices have changed so much since I started my career, I now feel bothered by the title of forester. What does this word convey now except exploitation? That is why I wanted to work in places other than Canada and Norway. Not only was I hoping to find that people elsewhere treated their forests with more respect and care, I also wanted to gain experience in different options regarding forest management. I wanted to study aspects of forest ecology, as well as native regeneration strategies in different kinds of forests and in different parts of the world. I wanted to discover if standard, north-temperate, forest-management thinking, as taught in North America and Europe, had worldwide applications.

History and cultural dependency on forests are easily forgotten. When I see yet another forest cover being lost this lack of awareness makes me cringe. It is like losing something of ourselves when forests disappear (Harrison, 1992). I am not arguing that we must stop logging—this is patently impossible—I am arguing that when we log primeval forests we lose something fundamentally important to our well-being and to our understanding of our place on earth.

Forests are also huge laboratories where nothing stays the same for long. Trees grow tall and age, new tree species get established, while others die and disappear. Natural disturbances, such as fires, floods, violent winds, insects, and diseases can change everything, and quickly. So much goes on in the forests of which humans get only glimpses. Forests are more than home to numerous animals and plants. They are communities that vary by habitat, soils, drainage, slope, aspect, altitude, and latitude, as well as by climate, rainfall, and history of disturbances. Interrelationships between plants, animals, and their environments are equally important. Remove one component, such as a tree, and all sorts of changes are felt throughout the system.

What we need is good forest management based on natural regeneration and supplementation by planting. We need government control of logging activities to support long-term objectives. If we want to have sustainable forests the profit motive cannot be the guiding principle. Foxes make poor protectors of chicken coops. By downsizing government departments that deal with lands and forests, governments are simply giving in to land speculation and providing industry easy access to our common forest property.

The many opportunities I have had to see and work in this earth's forests and with different people have not only broadened my view of my profession and taught me compassion for the world's poor, they have also filled me with despair at how we mismanage what we have been given. For example, places like the Philippine Islands (except for Mindanao Island) and Puerto Rico have lost nearly all their forest cover since the end of World War II. We have lost our sense of place in nature, and this alienation affects the management of all our natural resources. I know in my heart that I am richer for having seen so many kinds of forests that have such richness and beauty. I would like everyone to have such an opportunity — seeing might help us all develop more compassion and caring. My wealth of impressions around the world sometimes overwhelms me with feelings of outrage at current logging practices.

People have shared their ideas, knowledge, and opinions with me, and in doing so, have moved me deeply. Together we are a community that mourns forest removal because it is so often done in the name of dire human need. Do people know that when their indigenous forests are gone, their lives and ours will change forever?

I feel that I am among friends when I walk in any forest anywhere. That is more than many of us can say walking in cities. The disappearance of our primeval and "messy" forests is to be mourned for they connect us with a sense of reality and help people to be grounded in their lives. To write about forests has been a journey of self-discovery.

Pinus contorta latifolia
Lodgepole pine

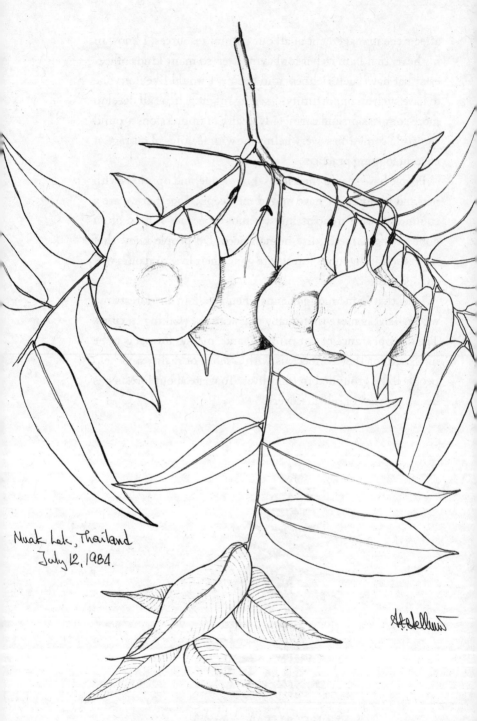

Muak Lek, Thailand
July 12, 1984.

PTERODCARPUS MACRODCARPUS Kurz.

A Place of Work and Wonder

Parents are sometimes dead set on deciding their children's futures. Mine suggested I study medicine, dentistry, or law. They suggested teaching and even philosophy—I cannot remember what else. It was their blessed understanding that because I could not make up my own mind quickly enough, they would have to do it for me. "Time," they said, "is of the essence. And you have so little time to decide."

Little did they know I already knew what I wanted to do but had not told them. I didn't want their interference. Besides, my mind was made up by the time I was seventeen years old and had worked my first stint as a forest labourer. To me forests were magical places, dark and safe yet mysterious and sometimes beautiful enough to invite silence and worship. Forests were places of wonder.

Few people I have met have known at that young age what they wanted to do for the rest of their lives. For me the realization began in my high school years (1949–1951) when our class, on several occasions, did community service. We worked on farms picking potatoes and cabbages, as well as in the forests north of Larvik, Norway. It was the forestry work that really appealed to me, but at that age, what was not to like?

Forests did not argue, push me around, or look over my shoulder all the time, trying to speed up my decision-making process. Even more appealing was that my parents had never thought of suggesting it. Had they, I might have hesitated. I did not know then how difficult and challenging a profession I had chosen.

Training Then and Now. Forestry is a practically oriented field of study and work. Without a good feel for the forest, it is easy to apply theoretical solutions to ecological problems, and that can lead to trouble in the long run. Mistakes are hard to correct and can take decades to put right. As forester Peder Braathe found in Norway, once you have started to reforest a temperate forest, where chance plays an even larger role, there isn't much you can do to alter its ensuing development (1953, 1976). We have one chance to reforest land after logging, only one chance to do the job right. In tropical forests nobody can foretell what is going to happen after logging because natural biological turnover is so tumultuous and competitive.

Between the 1950s and the early 1990s, people in Norway who wanted to study forestry had at least eight and a half years of learning ahead of them. First a candidate had to work as a labourer for at least two years. Then he or she could be admitted to study at a technical school for one and a half years. The candidate would then have to work as a technician for two to three years before being able to enter the three-year forestry program at the Agricultural College (now the Agricultural University of Norway) at Ås, a small community in southeastern Norway. For me, studying forestry under this program in Norway was not an option because only top students could ever hope to enter the program. My grades simply were not good enough. As a result, I cast my eyes on studies in Canada, a country that for a long time had held my interest as a land of opportunity. In the fall of 1955 I entered the forestry program at the University of British Columbia (UBC).

A lot has changed since then. In the early 1990s Norway reduced its total study requirements in forestry to five years, including only twelve weeks of practical training. Compare that to the four-year forestry program at UBC that gives students the opportunity to spend upwards of twenty-seven weeks on field work in the summers. Forestry studies today in North America and in Norway are therefore similar both in length and emphasis. The de-emphasis on practical training, however, is alarming.

Beginning My Professional Training. Having emphasized the need for dedication to practical professional training rather than just to an employer, I now want to describe why forestry work appealed to me so much in the 1950s and how it set the stage for my entire career.

During my last year and a half in Norway (after high school and military service, and before leaving for Canada), I again worked for the forest company in Larvik that I had in 1951 and 1952. That company, Treschow-Fritzøe, had large forest holdings north of my hometown, and I worked in their Løvås district. The forester in charge was a very enterprising fellow who made sure his trainees, called practicands, got as much varied work experience as possible. I was expected to work but also to challenge decisions I did not understand or agree with. That, I was told, is how you learn to make good forestry decisions.

During that time I worked as a logger, a scaler (measuring log volumes), a road worker, and a tree planter. I worked on weeding crews in forest plantations, blasted ditches to drain wet soils, and worked as a tree marker preparing areas for felling and for thinning dense stands of trees. During our field work we would have long discussions about what we were doing and why. The topic that came up time and time again was forest ecology. For example, practicands were asked why different species grew on different sites. Was this by chance or because different sites had different things to offer plants? Then we were asked how we might adapt our practices to fit these different sites. This meant understanding that each tree species has a specific set of site requirements. Pines prefer drier sites, larch need full light to grow well, spruce are shallow rooted, and so on. It was essential that we understand that plants, by their presence/absence and by their growth, revealed what any given site was capable of providing. Discussions such as these were always free and exploratory, and everyone was expected to contribute.

My Tool the Axe. I had to learn everything from scratch, including how to handle an axe and keep it sharp. The axe became part of my hand. I didn't feel dressed without its handle in my grip, and not because I was cutting down trees all the time, either. The axe had simply become my universal tool. I used it for splitting wood, opening tin cans, hammering, sharpening pencils, felling trees, cutting limbs off felled trees, and cutting blazes. I even once tried (unsuccessfully, I might add) to shave with my axe. Its blade was honed until it gleamed. Not surprisingly, no man could use another man's axe without first asking for permission. Even then the request was often refused or given grudgingly.

It should be easy to understand then how it became a matter of pride that a blaze on the tree should be perfectly cut and that the limb should be severed without damaging the bole (the trunk of the tree). The blade could not have any nicks or flaws in it, or it wouldn't leave a perfectly smooth cut on the face of the tree. Cutting into the wood or failing to remove all the bark and having to strike a second time, were common errors. The challenge was always to do better next time. Older field hands were there to teach me with patience and caring. The best blaze possible would just skim off the bark and let it hang down in a loose flap below the blaze. This way the blaze would appear twice as large. I had never before taken such pride in my work. Soon it became my ambition to become so skilled with the axe that the blaze would look as clean as if it had been made with a knife. Such a simple tool gave me such great pleasure. The axe became my badge.

Once when a chief forester came to visit where I was felling trees, he looked at the numbers I had cut into the logs and asked if I had used a knife. When I told him I had used my axe, I felt a flush of pride upon realizing what an impression I had made on my boss. I had achieved what I wanted.

The Way We Logged. Logging was all manual when I started to work at Løvås. Horses were used to drag logs from the forest to the riverside, and river transport of logs was commonly used to get wood to mill side. All trees to be felled were hand blazed and marked with the letter "L" (for Løvås) at chest height and just below where the faller should cut the stump. Logging was done by small clear-cuts, selective tree logging, and the thinning of dense stands.

The trees were felled using a Swede saw and an axe. At that time, in 1955, the chainsaw was just beginning to make an appearance. We continued to fell trees using a Swede saw and an axe. The trees were then bucked into standard lengths according to size and quality. Good, straight logs were used for lumber production. Whole tops, small trees, and curved or misshapen logs were used for pulp. All pulp logs had to be peeled by hand with a so-called barking spud. It was a labour-intensive task, which is why we were paid more for peeled logs than for sawn logs. Log ends were then trimmed to minimize breakage and splintering in the river drives. Each log, pulp, and sawn log was marked in half metres with a Roman numeral to indicate length.

The scaled logs were then dragged to riverside by horses. Here we stacked them in huge decks (some over thirty feet high), inserting sleeper logs between each layer so that the scaled logs would dry well and could be rolled easily into the river come springtime. The scaling itself was often done here at the log decks, rather than in the forest.

The felling work was heavy and very hard on my back. After using a Swede saw I could not straighten out for a while and would walk about like an old man. Adding insult to injury, I never made much money felling trees because I had trouble keeping my saw sharp and couldn't work as fast as the old hands for whom the rates were fixed. Fortunately this didn't make much difference because I still managed to make enough money to afford weekly food orders and some much-needed clothes. Housing was provided by my employer.

In the winter it was common to keep small fires burning all day in the area where we worked so we could keep warm when resting, eating our sandwiches, and for brewing pots of coffee. We were very careful to put out these fires at the end of every day.

During these times I remember how I'd stop now and then and listen. Listen to the wind in the trees. Listen for the sounds of birds and squirrels that were always around. It was a time for reflection. A time to breathe the fresh air spiked deeply with the smell of resin and to look at the work I had done. It was even a time to rest pitch-covered and sore hands that would stick to everything. These were special moments when I felt a tremendous closeness to the forests around me. My admiration and respect for them grew day by day.

In time I grew stronger and could even work alongside the old-timers without feeling clumsy or slow. I was also learning a whole new vocabulary of terms for our work. These were all fulfilling accomplishments. I started to live a new life of my own.

With all the effort I was putting into maintaining my tools, I couldn't help but think of my father's tool chest at home. His chisels were dull and chipped, his saws could hardly cut paper, and the heads on his hammers came off the shafts all the time. There was no doubt he was a good surgeon, but you certainly wouldn't know it by the way he kept his tools at home. Needless to say, developing a sense of pride in my equipment was a brand-new experience. I never really mastered sharpening my saw in the field, but I honed my axe daily and kept my barking spud in good shape so that it wouldn't skip on the wood and hurt my knuckles. Next on my growing list of new experiences was learning how to ski to work in the dark using cable bindings and rubber boots.

Roads were plowed often enough, but because there was rarely any transport available, I still had to ski to and from work, and often in the dark. When scaling timber on the slippery log decks it was necessary for us to "feel" the logs underfoot. That's why we used rubber boots. Leather-soled boots were very dangerous and could cause a person to stumble or slide off and get hurt.

Top: Trimming log ends at Løvås.
Bottom: Timber scalers and horses used to skid logs out of the forest.

The Last River Drive. In the spring of 1955, I also worked on the last river drive of logs in the Herland River. It was without a doubt one of the most memorable experiences of my life. One of my jobs included building massive log decks where the wood could dry during the winter and not rot or be stained by fungi. These decks were built near the river so that, come spring, the logs could be rolled into the riverbed as the water came surging down from the opened dam. It was exciting work but not without danger. The sounds of the logs grinding against rocks and each other, the occasional snapping and splintering of wood under pressure, and the force and wildness of the event were all dramatic. It probably caused severe ecological damage. Even though the river was little else than a mass of rocks and boulders, I hate to think what this did to the plant and animal life in it.

The river was not large so the dam water was essential for driving logs. Every available hand was needed. We made up teams of about six men and worked from dawn until dusk. This was spring breakup time. The snow was melting everywhere, and the days were getting distinctly longer.

The work was hard but our spirits were high and nothing seemed to vex us. Each morning started at dawn with someone going up the river to open the water gates. The rest of us rolled logs from the decks into the almost dry riverbed that awaited water. Preparing the riverbed this way allowed the greatest number of logs to be transported by the surge of water as it came by. After the water passed, around ten in the morning, we would rest and eat our morning food. Then some of us would follow the now almost dry riverbed, untangling piles of logs left behind by the surge of water and recovering those that had been washed up. The logs from the Herland River travelled into the much larger Numedalslågen River and then floated down to Larvik where the mills were located.

Top: Log decks by Herland River, spring 1955.
Bottom: Logs in the Herland River.

Some time in the afternoon, usually around two o'clock, we would again stop to rest and eat. Once our fires were lit for our coffee, we'd stretch out on the ground and munch our remaining sandwiches while teasing each other and telling stories. Other times we just lay down and slept on the moss until someone woke us. These were very congenial times.

Preparing the coffee was an event in and of itself. First we'd hang our tin cans over a stick stuck in the ground and positioned to lean over the fire. This stick was called, in Norwegian, *ei kjerring* ("a woman"). Then we watched until the water boiled, added some coffee, and waited for the grounds to settle a little. This was a man's world to be sure, and chewing on coffee grounds and spitting them out became a professional habit. Two cups a day was all I could afford to make. Every drop was poured and drunk with reverence. One worker would add more grounds to his pot until there was enough water for just one more cup. When his pot was full of grounds he would empty them out and start again. He said this saved on coffee, but we said he was stingy. No one wanted to taste his brew.

On some work days the dam would need to be closed to collect more melt-water for the next drive. This could take several days or even longer if the weather turned cold. At the end of such a day we would walk back to our cabins or if we were lucky, catch rides home on a tractor. There we would cook supper on our wood stove and clean up before finally collapsing into bed, happy and spent. Sleep had a sweetness all its own.

It was during this time that I realized how much I loved hard physical work. Between that satisfaction, the fresh air, and the comradeship often enjoyed in silence as we worked, I felt I had a place here in the forest, something I had never felt at home or in town. It was the first time in my life that I felt really accepted. I had a place among both men and trees. It was also during this time that I realized how much I loved the silence of the forest.

I could think of no place I'd rather be. In spring, summer, and fall, I'd jog along the trails in the mornings, the delicious smells wafting over me from forest to meadow to muskeg, each one with its distinct aroma. To get to work was a real pleasure.

A Passing Way of Life. When I think back to the events of that year, I realize what a privilege it was to be part of what was a dying way of life. We all knew it was the last river drive in the district and made a point to enjoy it. But everyone felt a twinge of sadness at time's passing. Trucks were replacing the rivers for transport, and the chainsaw was replacing the Swede saw. Noise came to the silent forest as trucks roared, belching diesel smoke, and saws could be heard for miles around. Before this time the forest's beauty had been accentuated by its silence.

These noisy new tools were also causing great disturbances. Noise isolated workers from each other. The faller often worked in continual fumes of oil and gasoline, and the tempo of logging increased. A sense of haste replaced the joy of work.

The power of machines outstripped both man and horse, and I knew that the communication between man and chainsaw was not—could not—be the same as between man and horse. The change was difficult to accept. Trees had been my friends when I worked with an axe and Swede saw. In a profound but simple way this work united me with the trees. Just as a hunter feels connected to the life it takes from a deer, so too did I feel connected to the life I took from a tree. I respected the trees for their size, beauty, and silence. I cared for them as one might care for one's garden. They were not just saw logs and pulp. They were stately, beautiful individuals, expressing all sorts of character. This feeling has never left me.

The Chainsaw Intrudes. Many of the old-timers I worked with scorned the chainsaw at first. They would pit themselves against them, trying to cut as much wood by Swede saw and axe as a man could with a chainsaw. But as time passed the old-timers had to give up the fight. But that's not all they lost. Before the chainsaw there was time to think about the job, the trees, and the environment. There was time to chat with the man and his horse that came to skid out logs. There was time enough to stop a while and listen to the sounds of the forest—time to straighten stiff backs, and to let sore hands slacken for a while. There was time to communicate with nature. But no more. As work pressures increased and noise followed wherever chainsaws were used, it became increasingly hard to listen.

The chainsaws and trucks used for log transport changed the forests, and they were no longer the majestic and silent places I had loved. It is why I have always considered noise to be our worst form of pollution. Even when one learns to live with noise it still isolates, robbing us of an intangible closeness to nature.

When I came to British Columbia in the fall of 1955 and witnessed coastal logging for the first time, I realized I had made a quantum leap into the future. High lead spars and donkey engines with cables, trucks that spewed black diesel fumes, and caterpillar tractors gouged the land—was this what Norwegian forestry would become, too? Everything about logging here showed an uncaring use of forests. It appalled me. The mess left behind, however, was "justified" because, as loggers told me, old growth trees were huge, requiring large equipment and roads, and old growth forests had a lot of rotten wood that had to be disposed of by burning. I had a hard time understanding and accepting all this. "Cherry-picking" was a new thing. Small outfits would come in after the major logging was finished and salvage small amounts of additional wood, but the mess and the noise all but destroyed my belief in forest management. Where was the management in all this? It looked like rape.

Fortunately, the following summer (my first in Canada), I was able to work in the pristine landscapes of the interior of British

Columbia. Here the sizes of the trees were smaller and comparable to what I was used to in Norway, and lodgepole pine and interior spruce were still frowned on as inferior to Douglas-fir. In the ensuing two summers I worked as a timber cruiser for the pulp mill at Hinton, Alberta, where horses were still being used to transport logs to roadside and mill. Horse logging here wasn't discontinued until 1968. These encounters with the familiar were reassuring, for by 1957 I was seriously beginning to question whether or not I had chosen the right profession.

I understood that mechanization was necessary to reduce logging costs and that much of the immediate site damage was temporary. But I also understood that mechanization was transforming the forest into something other than what it was before the noise, fumes, roads, and landings. There was more exploitation than I had dreamed possible. In Canada it was more important to get the logging over with than to protect sites from lasting damage. I became angry at myself for having compromised my professionalism so much just to keep my job, and I have often felt ashamed of how I failed to show forests the respect and care they deserved. Yet nothing dimmed my feelings of wonder and magic in the forests. In fact, their grandeur and silence still enchant me, even as my hair turns white and my limbs stiffen with age. Because I had the opportunity to roam and play in the forests around my home, and later work in them, I still feel a strong affinity with them today. That early sense of closeness has sustained me throughout my life.

Using the Forest as Sanctuary. The terror of World War II also influenced my childhood and made me aware of how silent and peaceful forests were. I was drawn to them when city and family fears frightened me. The forest environment stood in stark contrast to the turmoil, suspicion, fear, and discord that war and occupation imposed. Between the ages of seven and twelve

I lived in constant fear of losing everything dear to me: mother, father, and home. Father would many times be called into the Wehrmacht to answer questions. German officers came to inspect our house, intending to take it from us and use it as officers' quarters, and because mother remained an American citizen, she was very afraid of being sent to prison for being an alien. The forests around me were therefore also hiding places that I'd take long walks through as often as time permitted.

It was twenty years after the end of World War II before I stopped waking in the night from nightmares about persecution and danger. It was easy for me to understand why soldiers took refuge in forests. Not only were they hidden there, but they were also embraced—nurtured by their impartiality in matters of human enmity. It is also the reason why a Thai general could say in the 1980s that all forests along the Cambodian border should be felled to expose insurgents in time of war. It is why we must listen to forests and recognize what they represent. We must observe and work with them rather than destroy them in times of peace or war. We should listen from a position of humility, not from a domineering, controlling one, the way Westerners do. But humility is quite impossible now that we have the power to control and destroy forests the way we do today. Our respect for trees is disappearing as our control over them increases. Forests invite silence, not talk, noise, and fumes. In troubled times forests are a great comfort. In a most simple way, I can understand why Buddhist monks in Bhutan consult the forests before making decisions. The listening helps avoid hasty and ill-conceived actions.

Of course, my sense of safety within the forests was nurtured by the fact that Norway had virtually no dangerous, wild animals except for one type of poisonous snake. The environment was therefore safe, day and night, summer and winter. I never felt my life was in danger there. There were no robbers, no evaders of the law, and virtually no hermits, either. Instead, the forests were visited in droves by people hiking, skiing, camping, sightseeing, and gathering mushrooms and berries in season.

I cannot say the same for the safety of humans in Western Canadian forests. There, bears have chased, treed, and hunted me. On another occasion I was observed in silence by wolves at such a close range my hair stood on end once I realized what was happening. It was the reality of working in bear country, in places where little human development had taken place. This was their territory. I was the intruder. What I don't understand is why we plan recreational facilities, such as campsites and walking trails, without also controlling bear incursions. If we are to foster love and caring for trees and forests, we must provide safety as well. It is common knowledge that many city dwellers never camp or hike in our national parks and reserves because they are afraid of bears, cougars, and wolves. I even had to convince my wife it was safe to hike and travel with our children in the natural forests of Alberta. Later she gained familiarity and felt at ease.

The Kananaskis Valley in southwestern Alberta has been extensively developed for recreational pursuits and I have visited it many times to ski, hike, draw, and paint. I hesitate, however, to continue my visits because of a very close encounter with a grizzly bear some years ago. It happened in the Highwood Pass area. I had parked my car, walked into the forest, and sat down only to discover that I had forgotten something. I gathered up my materials and walked back towards the car. Not more than two minutes later a grizzly walked right by the place where I had been sitting. Had I been there, absorbed in my painting, and had the bear been surprised by my presence, she might have attacked. That is its natural instinct in such situations. When I got back to the car a game warden was standing there, watching the bear. "Isn't she cute," she said. Her ignorance frightened me. Did she really know what she was admitting? When I found out that someone had seeded the cutbanks of this upgraded road with, among other things, the seeds of a legume called Hedysarum, I realized that the right hand does not know what the left hand is doing. The root of this plant is a favourite food of grizzly bears. The plant had actually lured them to the very spot where people were most likely to stop.

As I see it, making Canadian forests safer for people would help save the forests from exploitation. Then more people might feel more comfortable venturing into them. As long as forests are thought of as potentially dangerous places we will have less compassion for them, less knowledge of them, and certainly more fear of them. Fear does not give rise to caring.

At Home and at Work in the Forest. Having grown up in a small coastal town in Norway, I often played, jogged, hiked, and skied in the forests. But I never truly knew forests until I started to work in them. I can't say I know a country simply by having visited it. I have to work in a new place before knowing anything about it, before feeling any communion with it.

In my many years of consulting abroad in forestry, I have shared my knowledge with people and tried to solve their problems. What I have found to be true is that urban people do not know their forests, tree species, plants, or animals. Some have knowledge, but only in ways that pertain to their own lives. The Arawak of Guyana, the people of Bhutan, and my fellow workers in Norway are the only people I have worked with who knew their forests and felt at home in them. They could identify tree species and other plants by their local names, as well as medicinal plants and those used for various local needs. They also knew about animals because they depended on them in a way that city people never do. It is surely the same for Aboriginal people of Canada. However, city people are also generally the same everywhere. The Blacks and the East Indians who lived in the cities adjoining countryside in Guyana, for example, were as unaware of their forests as are the people in Edmonton or Calgary. How often have I heard from Edmontonians that all boreal conifer trees are "pines." It doesn't matter whether the tree is a spruce, a pine, a larch, a true fir, or a Douglas-fir. They are nevertheless called "pines." Very few people know the difference between jack pine, lodgepole pine, whitebark pine, and limber pine. Yet these are all species native to Alberta. By the same token, poplars and aspens are lumped together as poplars by some and as aspens by others. I find it neither surprising nor upsetting. It is simply the way it is.

Listening to Trees. I started to ascribe human characteristics to trees when I worked at Løvås, Norway, for the better part of two years. No, I am not crazy. It was similar to what hostages may feel in captivity. They may start to take the side of their captors and feel anger at the society they came from. It is a similar experience working in the forest. Forests and trees are my lifelong friends, and I am protective of them. I argue that it is this feeling of affinity that we as foresters need to redevelop to be good custodians of this resource.

When I am in the forest I am aware of the personalities of different species of trees. Pines, for example, wave their heavy branches as if in welcome to the weary. Hugh Brody says as much in his book The Other Side of Eden (2000). Pines wave as if they would enfold and comfort. I suspect many readers will laugh at my subjectivity. But I know I am not alone with these feelings.

I also feel that spruce trees, in their rigid majesty, are like judges that protect the forest domain from intruders. They also bring darkness with them as they grow older and shake their boughs stiffly in the wind as if they had just come out of a shower. At times they remind me of judges wearing slightly messy wigs. They are the silent majority in older forests of Alberta. The spruces are austere and dark while pines are open, welcoming, and friendly.

Birch and larch forests, on the other hand, represent light and youth and are, therefore, forever graceful. Aspens are like chameleons, changing from season to season. Using feminine gestures they strut their colours in spring and fall with a hint of vanity and exude perfume in springtime when their buds break open. In the least breeze their leaves tremble nervously, but they can also be opportunists, moving quickly into disturbed habitats when the chance arises.

Then there are the poplars that remind me of the newly rich—unkempt, large, and fast growing, flaunting their power through pure size, even though their life spans are normally short.

To me, these characteristics make up our western prairie and boreal forests. They give each place a character of its own, varying continually, depending on circumstances. Maybe if we all felt this way about trees, we would treat them with greater respect and think of them as more than things to harvest. To do less is an insult to their perseverance.

When the ice melted after the last ice age, some nine to eleven thousand years ago, it melted in a northeasterly direction in Alberta, taking two thousand years to disappear. As bare ground became available, willows, poplars, and aspens shed their seeds onto moist mineral soil, germinated, and established themselves. But since that time bare mineral soil surfaces have been invaded and occupied by other vegetation. Today, aspens in Alberta rarely produce seedlings from seeds and instead propagate themselves vegetatively. It is therefore entirely possible that some of the aspen clones in Alberta could be thousands of years old. How could we not admire and respect such venerable age and tenacity?

If we are to protect our forests from destruction and exploitation, and if we are to manage them sustainably, we need to feel in our hearts that forests are living ecosystems with innumerable, interdependent individuals living within them. We need to have confidence in their restorative powers, working with them rather than forcing our ideas on them.

Not Listening to Trees — or People. We don't listen to trees when we plant where natural regeneration is in the process of establishing itself. By planting right away we ignore what the forests can do for themselves. Before we plant we should wait to see if stocking will be adequate. Some planting is obviously necessary but natural regeneration should be considered first and foremost and planting should only be a second, fill-in option when natural regeneration fails. It isn't that I don't believe in planting as a technique or in genetic improvement. It's that I firmly believe that Mother Nature can establish seedlings better than we can. I have seen enough bad planting projects to know. I've seen rigid utilization standards imposed on forests, with trees planted in rows or in patterns governed not by Nature but by what site preparation equipment dictates. This is no way for a forest to be structured. Trees cannot grow everywhere, in every microhabitat.

Man-made, planted forests cannot ever be made to look like natural forests; they will never have such large trees again, or ever be as diverse. They cannot be made to look like undisturbed places or like places of sanctuary and beauty the way primeval forests can. That would go against the grain of forest management, where regulated growth is the aim.

When we destroy these habitats, often without thinking or knowing, we harm not only plants but also endangered animals. As our protected or "wild" areas grow smaller, habitat preservation becomes harder to carry out. Foresters who know what is happening are often tongue-tied and keep their silence.

I remember going to a meeting once in Georgetown, Guyana, where a logging company manager was explaining his company's plans to the interested public. He brought a blueprint to indicate the Arawak villages in the area in which logging would occur. There was a tight circle drawn around each of them, and the company manager referred to them as "protected." I presume he meant that the company did not intend to fell trees in the villages themselves. That was all. If anything, the villages would be doomed, not protected. I knew by the plan that the people would lose traditional hunting grounds for years after logging ended because animals would move away for a time. People would be without the forest cover they relied on for fruits, nuts, plants, animals, and protection from the sun. Their places of domicile would become hotter, drier, and far less comfortable. Even the building materials used for their houses might be gone. Sadly, I was the only one who knew this because not one Arawak was at this meeting. They hadn't even been told of it. The lack of caring and understanding on the part of the logging company made me angry. There was no one there from the affected community to present the Arawak's case. When I voiced concern over this I was simply ignored.

As a European by birth I cannot escape the feeling that forests everywhere are seen mostly as having potential profit. Utilization standards change all the time, and twenty-year economic planning horizons are too short for managing forests over periods half a century or more. Then there is this inflated thinking that by replacing a cut and taking credit for the natural regeneration that supplements our partially failed plantings we have done our jobs. We talk a lot about sustainable forestry but in much of the forty per cent of Canada's forests where logging is carried out we will not see commercially valuable forestry practiced again for a century, if ever.

Conclusions. Forestry practices have changed completely since the forties, when manual labour and horses dominated forestry work. Because of mechanization we think bigger than before and harvest forests so rapidly that some cannot survive the pressures. Forests have become so valuable that speculation and recreational developments are threatening even the most remote parts of what used to be untouched forest domains. There is hardly a forest anywhere that is safe from human invasion.

My concerns embrace many aspects of forestry, including training. Today, for example, we downplay the importance of practical training in our universities in Europe and North America. University graduates now enter the work force to start their practical training as employees with hardly any knowledge of what it means to practice forestry in the field. At least in North America, teachers and professors often have little or no practical training behind them and can therefore not speak from experience. Students miss this grounding. They learn far more theory than practice. As employees, their on-the-job training is limited to one or two employers.

Complicating matters, we in the West have so distanced ourselves, and continue to distance ourselves, from the forests that we no longer know them as people did just a generation ago. In fact, our fear of forests increases as our lack of awareness of them increases. Bugeja (2005) expresses particular concern for the technology that leads us into living what he calls "virtual lives," dealing with all manner of issues, including nature, via remote control.

A consequence of living this way is that we are unable to remember what the land and forests looked like fifty years, or twenty, or even ten years ago and, therefore, are willing to accept whatever the last generation did to our forests. We don't understand that we are accepting and living with increasingly smaller and less diversified environments than we lived in just a decade or two ago.

Although my concerns embrace many aspects of forestry, I find it particularly disturbing that we have shortened our training requirements for practicing forestry, as well as that for practical experience. In our universities in Europe and North America, practical training now comprises only modest parts of the curricula. These changes, and others, make clear that most foresters practice armchair forestry from boardrooms and offices far removed from logging. More than ever industry now trains foresters on the job. Academic training has become more and more theoretical and less and less practical. Young university graduates become faithful employees and cannot, or do not, speak out against unprofessional logging and management practices. No wonder the public considers foresters part of the problem rather than part of the guarantee that forest management will be carried out sustainably. Foresters may think they are professionals, but as far as I can see the fact that they cannot speak freely tends to make them technicians rather than professionals. I keep looking for public discussion about forestry matters. The silence is audible, but it is not the wonderful silence of the forest.

As we practice forestry more and more by remote control, and predominantly with logging in mind, we run the terrible risk of damaging our forest ecosystems with our single mindedness. By thinking big, attention to details suffers. If we want to maintain biodiversity in our forests, we have to pay attention to details. Forests are made up of thousands of individual plants and animals that have their own requirements for nutrients, light, water, warmth, and cold.

More important than anything else, our populations are increasing almost exponentially and our need for space, cropland, and wood is outstripping what our world can sustainably provide. Until we come to grips, worldwide, with this problem of overpopulation, forests will continue to disappear at somewhere between 11.3 and 15 million hectares a year (FAO, 1986; Williams, 2003), or more as our needs increase.

We are facing the extinction of primeval forests on a worldwide scale. The mysticism, protective nature, grandeur, and power of primeval forests are all disappearing as we push computer buttons and plan timber harvests. In losing these forests we lose something of our human roots. Our loss of bonding with nature is catastrophic, and I see no end to this delusional belief that we can manage forests better because of computer simulation, data banks, and theory.

I am glad I had the chance to live with and learn from trees when I did. I also wonder how young foresters today see this world and perceive how to interact with it.

Larix laricina
Tamarack

Picea abies
   "Norway spruce"
Eggedal, Norway
June 29. 2005

Forests are Living Ecosystems

A tree is more than a source of wood or beauty. A forest is more than a hinterland where dangers lurk and people get lost. Trees are living members of communities, just as you and I are members of the communities where we live and work. Both depend on their communities for their livelihoods. That is why it is terrible that many people see trees only as sources of potential profit and fibre.

A typical example of this dependency in forest communities is called "wind throw." Once forests have been opened through logging or natural disasters, trees along the exposed margins will often fall because their neighbours are gone. The mutual support on which they relied to support them is disturbed. The same applies to human society when any of our support systems fail.

If we say that we know how forests should be logged, regenerated, and managed sustainably, we are telling only a partial truth. If "wind throw" occurs it often signals a misjudgement of soil-moisture conditions which causes shallow rooting, or misjudgement of species' characteristics. We know that nutrients, temperature, light, and water affect growth, but we rarely ask about the biology of inter-relationships between trees, other plants, animals, and micro-organisms. If we are to manage trees adequately, it is important to think holistically about them and their environments. This task is, however, very difficult. We don't understand how important all the components of an ecosystem are, but we are trying to find markers that might guide us, including the presence or absence of arthropods (Langor et al., 2006a) and rotten wood on the forest floor (Langor et al., 2006b). When regenerating forests after logging, we have only one chance to do the job right, according to Braathe (1953, 1976). If we fail, there is little we can do in the short run to correct failures. This one chance puts great demands on us to do the job well the first time.

The loss of forests changes environments for both tree species and humans, at least temporarily and sometimes forever. Logging can change a tree's ability to regenerate itself. If a tree is shade-tolerant, for example, it has trouble establishing itself in open, logged areas where light and heat may be more intense than the seedlings can tolerate. It is probably safe to say that most of the world's economically valuable indigenous forests are filled with trees that tolerate or require shade. We still log old-growth forests, but we are replacing these forests with species that can grow in the open areas we create by logging. We are, therefore, changing the very nature of our forests. The white spruce, for example, usually comes in under the shade of other trees and prefers only about thirty-five per cent of full sunlight for best growth. The lodgepole pine, on the other hand, which is a shade-intolerant, sun-loving tree species, usually establishes itself in open areas under full sunlight after fire and logging. It is often the pine, therefore, that provides the shade for the later spruce. But if we cannot wait for plant succession to take place, what then?

The webs of forest communities are every bit as complex as human communities. Too often forests are thought to consist mainly of trees; however, forests are much more complex ecosystems. They do not make up the ecosystem by themselves. Insects distribute pollen and can help seed germination. Birds eat tree seeds and distribute them by defecating where the seeds might otherwise have never been disseminated. The stomach acids in the birds also scarify the seed coat, which then makes germination possible. In other parts of the ecosystem, fungi infect tree roots and provide nitrogen; trees support each other in heavy winds and provide shade for other species. Trees also provide homes and nourishment for birds and other animals.

In 1984 a study in Thailand revealed that the seeds coats of a leguminous tree called Sesbania were so hard that it took years for them to absorb water and germinate. The seed coats literally had to rot before germination could take place (Hellum and Sullivan, 1990).

Fortunately, there is a wasp in the Muak Lek area of Thailand that lays its eggs in the seeds of this species as they start to develop—well before the seed coats have become hard. When the new adults of these wasps chew their way through the hard seed coats and emerge, many seeds can suddenly germinate and do so quickly. As a result, the seeds can absorb water through the exit holes left by the emerging insects. The seeds we studied were heavily damaged by this wasp. Up to ninety per cent were infested. Even though the insect larvae had fed on the embryos and cotyledons inside the seed, fifty per cent of the infested seeds could still germinate and did so within only two or three days. This kind of symbiosis is not uncommon between plants, animals, and insects.

By contrast, man-made forests lack diversity and cannot easily be made to look like indigenous forests. In fact, foresters try to avoid complexity in the name of efficiency. Planted forests, therefore, are susceptible to losing most of their diversity of animals and plants, which is especially true when exotic species are introduced into new environments. Such new species may be of no value whatsoever to native animals or birds. Radiata pine plantations in both Australia and New Zealand are proven examples.

I am, of course, not listening to the claims of vested interests who say they know how to log sustainably and to regenerate successfully. I know that the further we move away from the temperate forests towards hotter climates the more false this claim becomes. An example of this occurred in Brunei when local foresters asked for help in 1986. Some Japanese businessmen had offered to log their very special sal* forests for a good profit.

---

* **Shorea albida.** Mabberley (1987) uses the vernacular name of sal for one species of the genus **Shorea (robusta)** only. I use the name sal to cover species in both the **Shorea** and **Dipterocarpus** genera of the family Dipterocarpaceae, to conform with the common usage of the name sal as I have found it in Southeast Asia.

These forests grew in mangrove-like places on deep deposits of humus. They were swamp forests. Was it true, the foresters asked us, that there were people who knew how to regenerate these forests artificially? They themselves had tried and failed enough times to doubt such claims, but the Japanese men insisted it could be done.

The Japanese helped log most of the sal forests north of Manila, on Luzon Island in the Philippines. However, after seeing their efforts at reforestation in a place called Karanglan, I realized the foresters didn't know (or care) that you cannot plant shade-loving sal species in hot open places and expect much success. The landscape was scorched and the seedlings, planted again and again, wilted and died in succession. Even rudimentary knowledge of forest ecology should have told these earnest people that such efforts would come to naught. Because of this experience I was sceptical of the claims of success the businessmen were making in Brunei. No matter how valuable these earlier sal forests had been at Karanglan, these areas should have been regenerated with other shade-intolerant species to protect the sites for later sal plantings.

Worldwide, forests are mostly in the public domain, managed by governments and owned by the country's citizens. But on numerous occasions the view has been presented that private, rather than public, ownership will lead to far better forest management, simply because private ownership implies a closely vested interest in sustainable timber supply. This kind of thinking fails to take into account the fact that forests are far more than wood factories. They are resources that humans cannot do without. Therefore, selling them to private interests should never be an option. Human societies have succumbed when forests have been lost to over-exploitation or climate change (Fagan, 2004; Williams, 2003). One doesn't have to look only at small island cultures to see this truth, even though the fate of Easter Islanders in the past illustrates the point most vividly (Diamond, 2005). Easter Islanders logged every tree, and then turned to cannibalism to survive. This world is as finite as Easter Island, larger, but still limited. Fortunately, we have more time to do the right thing by Mother Nature than the Easter Islanders had.

Even when forest land is not lost altogether to other uses, its productivity is often damaged by over-exploitation. When Guyana gained independence from Britain in 1966, the so-called White Sands forests were repeatedly cut for charcoal production and timber to supply domestic needs at a time when people could not afford imports. This belt of near-coastal forest is now virtually useless and so hot and dry that new forest growth is at a standstill. Had foresters been alert and had they been mandated to control and regulate this use, perhaps these White Sands areas would still be productive today. Instead, the area is a wasteland— a place where people moved in, burned charcoal, and were never stopped.

To ignorantly downsize government departments in areas of natural resource management is dangerous for our well-being and our future (Wood, 2004). For example, government foresters in many areas of Alberta used to carry out their own reforestation after logging. They, therefore, had the experience to set guidelines concerning logging and reforestation, but because they no longer have to carry out reforestation they have lost the knowledge needed to do their jobs well. As a result, foresters trained in theory—not practice—decided how fast seedlings should grow to meet production targets; they failed because they over-estimated growth rates. Having never understood how fast natural seedlings can grow in local conditions, they did not understand that their plantations would not meet standards of restocking without a natural ingress of seedlings from seed. This one mistake cost the industry and the province millions of dollars and considerably increased the cost of regeneration practices.

Bad Days. Forests are there for the taking. As long as vested interests (such as dispossessed and hungry people or logging companies) are given easy access to land, forests will disappear. That is why foresters need to tell people why and how trees, such as those growing in the coastal White Sands area of Guyana, came to grow in dry areas. Questions should have been asked. How long did growth take? Why weren't soil moisture problems anticipated in such sandy soils? Potential users needed to know how long it takes for a succession of plants to get established, and to realize how long it would take to re-establish a second forest there, should our logging be too severe. The answer is centuries. To avoid such consequences we need to establish how much we can log from stands of trees to avoid over-exposing them to the glare of the sun and making the habitat too hot and dry. From our studies it became clear that nothing less than careful selection logging would be possible in white sands areas. In contrast, north temperate forest soils tend to be too cold for good forest growth. Because logging opens these areas, the soils warm up, which improves the growing conditions for tree seedlings. This is more proof that forestry practices must be site specific.

Site degradation, both aesthetic and physical, follows on the heels of poor forest management. When the White Sands area of Guyana was clear-felled, soil-surface temperatures during the hotter parts of the day increased dramatically, approaching 80°C in the early afternoons — hot enough to kill seeds and stunt growth. Sandy soils with very poor water-holding capacities also severely hampered growth there. When I went into these scrublands at daybreak I found a great amount of dew and guttation on the leaves of most of the plants there. My guess is that this might be the main source of water for young plants during the day.

For forests to return in denuded areas they need to be left alone for a while. In Bhutan it took a human catastrophe to give the forests a chance to reclaim the land they had previously occupied. In 1990, when I looked at the young forty- to fifty-year-old blue-pine forests growing in northwestern Bhutan (particularly those in the Thimphu and Phubjikha valleys), they seemed out of place. I remember wondering why these young, even-aged forests were growing adjacent to human settlement. They certainly had not been planted there. There was something decidedly artificial about them. Something must have happened in the past to permit this reinvasion. Someone then told me that maybe forty to fifty years earlier many people had died as a result of an epidemic. What followed was a chain reaction. Once people died, cattle died; once grazing pressures were eased, the forests could come back. That was the reason blue-pines were growing there in the 1980s and 1990s. Today these very same forests are again under severe pressure from both cattle and humans for fuel, fodder, leaf litter for the paddy-fields, apple orchards and building sites. This tussle between humans and their cattle on one hand and the forests on the other is a familiar one. The pressures on these forests were once again increasing, but people didn't seem to be aware of what was happening. History was repeating itself.

Another problem that humans have is the lack of perception of change. We are notoriously poor at perceiving changes that come gradually. I cannot remember what the trees on my property looked like when I moved onto it thirty-five years ago, any more than the people in Bhutan see that forests gradually give way. I have to resort to photographs to help me remember. We see only what is, are too willing to accept what past logging and land-use practices have left behind, and think that what we have is wilderness.

The Introduction of Foreign or Exotic Trees. To have an exotic plant in our garden is like wearing a new dress or suit to a party and being admired for it. It is nice to have something that no one else has. Our forests are changing gradually as we manipulate their growth through breeding strategies and the use of exotic species. As we introduce more and more different species and strains they are taking on different colours and shapes. Experimentation with different species has been ongoing for millennia. To be able to grow an exotic tree species far away from its origins is fun. Thinking of the plant's well-being doesn't cross our minds, even though plant form, size, and beauty all determine how much such plants are worth and whether or not they achieve their biological potential. Darwin's The Domestication of Plants was published in 1868. He expounded on how many generations it would take for an exotic species to become adapted to a new home and climate through natural selection, provided that the species was able to reproduce by seed. He was thinking of agricultural crops, not trees, and he thought that adaptation, of sorts, might take place in three or four generations. If this same thinking applies to trees we are never going to find the answers because three or four generations for trees could mean more than five human generations. It takes that long for some trees to reach sexual maturity.

Moving plants around the globe has been a human pastime for probably a millennia. In fact, most of the domesticated crop plants we use today have been imported from somewhere else, but we are just now starting to experiment with trees in the same way. We haven't always been good stewards of these imports, even if we have been able to keep many exotics alive in most unlikely places, far removed from where many of the plants originated. The Linnean Gardens at Uppsala in Sweden is a case in point. Darwin was addressing the need for colonial officers to grow their own food in the furthest reaches of the British Empire. People were getting tired of exotic foods in exotic places and wanted to taste dishes they were accustomed to at home. When it comes to food crops, we have obviously introduced exotic species into new environments.

But trees take a very long time to mature. Moving them around is a greater problem because mistakes show up only decades, if not centuries, later. We can experiment on a small scale, in private gardens, nurseries, and arboreta, but today we plant whole forests with strange species. The consequences of such ill-advised attempts are far reaching in both time and space.

The Swedes, for example, introduced lodgepole pine from northern Alberta and the Yukon only to discover years later that they had made a mistake. When they collected lodgepole pine seeds from trees around the sixtieth parallel in Canada and planted the trees several degrees further north in Sweden, where the climate was similar to that of northern Alberta and the Yukon, the trees grew thirty per cent faster than their local pine trees. Many years later they discovered that these trees fell over in strong winds because they had not produced a proportionately bigger root system to support their above-ground trunks and crowns. Planting programs with lodgepole pine in Sweden have since been severely curtailed, but the trees that had already been planted will create problems for years to come. Every time a strong wind blows more trees will blow down and more clean-up will be required.

The Malaysians introduced an acacia species from Australia (Acacia mangium) with much the same result. Acacia species are known for their copious seed production. Once introduced they are very difficult to get rid of. Introduced species sometimes become so successful in new environments that they out-compete local vegetation and become serious pests.

Another classic example is the Ipil-Ipil tree (Leucaena leucocephala) from Central America that was introduced worldwide for its fodder value, fast early growth, timber, and its ability to fix nitrogen from the air. Unfortunately, an insect called psyllid has infected these trees around the world and has devastated many plantations.

Trees are moved around the globe to meet many different needs, but forethought and testing are often inadequate. To foretell the possible damage from insects or disease is impossible, so failures occur. I don't have to look far to see that commercial nurseries, in spite of growth-zone information, try to sell untested outdoor plants that are not at all adapted to the climate for which they are intended. There is such a nursery in my immediate neighbourhood. We simply don't take the time to learn if the introduction of a new species is a good thing before we launch large-scale projects. This is especially true in reclamation work where sites have been so disturbed that aggressively colonizing species are needed. Such immigrants can turn out to be real problems rather than assets in their new homes.

Sometimes, however, trees planted in new places are valuable additions to nature's flora. Plants develop relationships with their indigenous neighbours over millennia. When they are suddenly moved to places where they don't have the same checks and balances they may develop unexpected traits. For example, eucalyptus species often grow much faster and better, exhibiting completely different growth forms, outside of Australia than in their native land. Radiata pine, planted in Mediterranean climates, grows much better there than it does on its native Californian Monterrey Peninsula.

Northern hemisphere conifers have been planted extensively in Africa, Australia, and New Zealand, and many other tropical countries to provide long-fibred wood for both papermaking and construction purposes. India and other countries have imported eucalypts and acacias on a large scale from Australia. These trees are used to revegetate denuded and dry habitats which were often created through misuse of the land in the first place. They also furnish much-needed wood. Australian acacias have also been planted in the Cape region of South Africa to stabilize shifting sand dunes that posed a threat to adjacent vegetation.

The introduced acacias have reportedly been effective in stabilizing the Cape dunes, but they have also adversely replaced local flora, aggressively spreading their domain through profuse seed production. These acacias were purposefully brought in because they grow in dry places, but now that they have done the job they are becoming a nuisance. Species that grow in dry areas often grow much better in moister habitats where they may not be wanted.

Being Able to Identify Tree Seedlings. Collectors of plants in the past couldn't send home seedlings of the species they were collecting because they didn't know which seedling represented what species of tree. To learn which seedlings belong to what tree takes time. As a result, even today, herbaria all over the world lack examples of tree seedlings.

Because foresters or forest workers often cannot identify their tree seedlings either, they cannot predict what a new forest will look like after logging or natural catastrophe. For the same reason, people cannot tell what the new forests will look like, nor can they save good natural tree seedlings in brush control operations. People simply cannot distinguish between a good tree seedling and a weed. I worked in one place where ten perfectly good natural tree seedlings were killed for every runty little planted seedling that was saved because no one even bothered to find out if the area was fully stocked with natural regeneration before planting was undertaken. They needed botanists but didn't bother to solicit their help.

The dynamics of forest regeneration become increasingly complex and ever more variable as one approaches the equator. By and large, familiarity with tree seedlings is not a problem in temperate forests because species diversity is not so great that identification causes problems. But in the tropics, it is an almost insurmountable problem. Here one really has to know how to identify hundreds of different seedlings to tell what a logged area will look like in the future. It is not enough to know how to identify tree seedlings because one has to know shrubs, vines, and herbs, in their seedling stages, to be able to recognize that trees are not just brush becoming established.

This problem of identification is exacerbated because tropical forests regenerate themselves in different ways than do temperate forests. In tropical forests, species come and species go because conditions for establishment there can be so severe that timing is all-important. In temperate forests, on the other hand, individual seedlings of a given species may die in a plantation but that species does not usually disappear altogether.

The lack of availability of local tree seed and problems with species identification are valid reasons for foresters to decide to use exotic species instead of local species in reforestation work. Besides, forest nurseries don't often have planting stock of native species. They often cater to species that are exotics. Seed collection is a problem when seed years are infrequent and when trees from which seeds are collected may be widely scattered, making collection both difficult and time-consuming.

This problem of tree seedling identification has made me embark on a personal quest. Wherever I have worked overseas I have tried to publish handbooks to illustrate what selected tree seedlings look like (Bhutan, Tshering and Hellum, 1990; Guyana, Hellum, 1994; Thailand, Hellum et al., 2000; Malaysia, Hellum et al., 2005). The task is enormously large and beyond my capacity to address fully, except to offer examples of what is needed.

We think we can solve forestry problems, or poor land-use problems, by introducing miracle trees, but there are no short-term fixes in forestry. When we introduce exotic species into new environments, we do so to solve problems. If we are successful in introducing a given species that solves our immediate problem, we will, most assuredly, be left with a larger problem. Therefore, we have to think one step further than we do today. We have to think of what happens after we have become successful with what we are introducing. For instance, we may not be able to get rid of an aggressive competitor once it has done its job for us. This is true of the acacias in the Cape region and in peninsular Malaysia, for the eucalypts near many Indian villages, and perhaps even the pines in Australia and New Zealand. Certainly the plantations of pine I saw in Guyana were less than successful, being almost completely overtaken by local vegetation. All that had been accomplished with planting pines in the first place is to have set back natural regeneration by a good thirty years.

If the relationships between trees and their environment are not properly understood, forests cannot be sustainably managed. They will certainly not be managed ecologically, and local people may cope a long time with undesirable species. In the process we homogenize the vegetation of this world. I heard a forester once say that he had entered his profession because he was going to be dead and long gone by the time his mistakes were discovered. Foresters don't live long enough to learn from or to correct their mistakes.

Some people firmly believe that plantation forestry using exotics is here to stay (Evans, 1985). They argue that there are enough success stories to justify this philosophy and practice. Evans' research lists several successful examples of this, especially in Africa. He does not dwell on failures but instead writes about operational problems. Readers are led to believe that operational problems can always be overcome.

Trees take a very long time to mature. They can take more than three, four, or more human generations to reach their own maturity, by which time our uses of those forests may have changed, or our grandchildren may have taken over and changed everything we started. I shall never forget the Norwegian forester who took me to see an old stand of Norway spruce trees back in the 1950s. It had been thinned and pruned for the better part of a century to produce top quality lumber. "This forest," the forester told me with sadness, "is now worth more for pulp than for lumber. It will soon be cut and ground up for that purpose." What can we do in forestry today to foresee future trends and needs? How can we ever do the right thing?

Changing Environments. Jared Diamond (2005) in his book Collapse talks about how changes in climate, brought on by human interventions or otherwise, have influenced human settlements. He describes in detail how the Norse settlers in Greenland succumbed to the effects of climate change because their cattle and sheep overgrazed the dwindling pastures as the climate deteriorated. The Norse were also unable or unwilling to adopt Aboriginal ways of coping by changing their eating habits and wouldn't eat fish of any kind. Instead, they ate meat and only resorted to seal at the very end, when it was too late. We are doing the same thing to our forests today—overusing them worldwide, even though we know our actions are not sustainable. The Norse settlement eventually disappeared because the settlers could not, or would not, adapt to a changing environment. They persisted in overusing their resources to maintain their traditional lifestyle. Is this not what we are doing to our world's forests today? Even though we may realize that we are wrong, many vested interests persist in carving up forests because society lets them carve away. In Malaysia in 2003, I talked to a sawmill operator who told me that he would have large wood to cut for no more than thirty years at best. That would be the end of his operations. In his book, Diamond shows clearly that people have a choice. Some people survive and some do not. For forests, it often depends on stewardship, on how well we listen to trees and to what is really happening to our world's many environments, forest environments included.

Williams (2003) stated that up to fifteen million hectares of forest land are lost annually, worldwide. This total is twice the forest land-base of Norway. Twenty per cent of Norway is covered in forests. I read once that it took 17,000 mature conifer trees to produce one Sunday edition of The New York Times. That is roughly equivalent to clear-cutting nearly twenty-eight hectares of forest in Alberta for this one event. The questions are always how long can this go on and how long do we have to figure it out?

The loss of forest land to other uses is especially alarming in Southeast Asia. It is reported that Thailand lost half its forest land base between 1960 and 1980 (Wasuwanich, 1984), and even in the 1980s it was losing over 550,000 hectares of forest annually to other uses, judging by remote sensing evaluations (Klamkamsorn and Charuppat, 1983). Thailand is now a net importer of wood. But the stories of forest removal come from everywhere. Haiti has lost almost all of its forests while the Dominican Republic, just next door, has preserved much of its own (Diamond, 2005). Puerto Rico had lost ninety-five per cent of its forest cover to subsistence farming by 1984 (Smith, 1984) and Brazil reports that it lost 12.8 million hectares of Amazonian forest between 1980 and 1995 (Cattaeno, 2002), which is an annual loss of over 850,000 hectares over this period.

The world is also becoming an increasingly artificial place, manipulated and maintained by humans and their chemicals. Population pressures drive this trend. What happened in Greenland many centuries ago may happen to this entire world, perhaps in the foreseeable future. Many think that human needs are the only needs that matter (Trefil, 2004), but there are types of ecosystems, forests, and oceans that require great care and attention because we depend on them for food. If our forests were used sustainably and looked after as we do our financial investments, they would be protected and watched as they grow rather than allowed to dwindle.

Again and again we are told that we live in a period of global warming that is accelerated by the pollutants we put into the air. We are told that we need to curb our way of using biological resources in order to moderate the speed with which our climates are changing. Not only must pollution be curbed but so must our rapacious use of this planet's biological resources be reduced. By putting too much carbon dioxide into the air we are heating up the planet and melting polar ice caps more quickly than might happen naturally. When more and more fresh water enters the oceans from melting ice, ocean currents can change. Without the Gulf Stream, Europe will be far less habitable than it is today. When the large amounts of melt-water trapped by the ice in Lake Agassis, at the end of the last glacial period, drained into Hudson's Bay and then into the North Atlantic, the event caused a mini ice age in Europe (Diamond, 2005).

We are told that forest removal implies more carbon dioxide in the air because the forests that use the carbon dioxide are vanishing. And we are told that this could lead to the earth heating up. There is little difference between the people who know this and the people who contribute to the problems. All are consumers, including you and me.

History has repeated itself over and over in the past million years, during which there have been four major glacial periods. These periods were not aggravated by humans polluting the atmosphere but by natural instabilities in the way the earth functions and in changes in the sun's behaviour. But now changes are happening at an alarming rate, faster than past trends of change would have predicted. Surely there can be no doubt that we are to a large degree culpable, no matter how unpleasant this fact may be.

Picea glauca
White spruce

Picea mariana
Black spruce

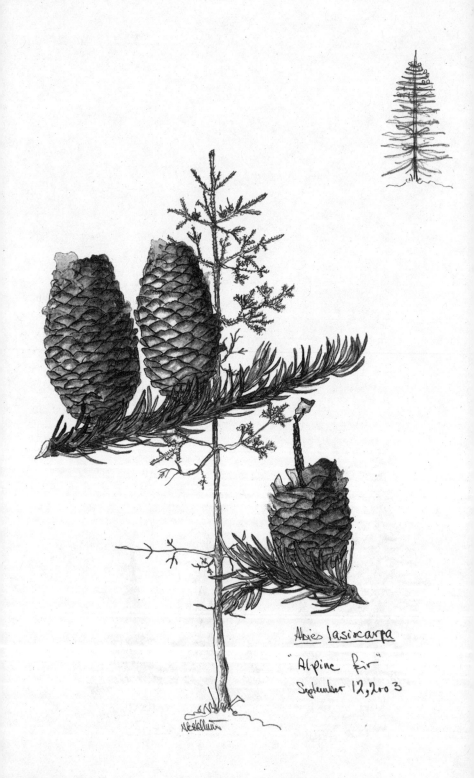

Abies lasiocarpa
"Alpine fir"
September 12, 2003

Examples of Failures and Successes

Examples of forestry failures, especially in tropical forestry, are plentiful. Examples of success are fewer. In 1986 I once walked a whole day in South Vietnam just to see one small but good stand of naturally regenerated sal trees. This, of course, was after the terrible deforestation caused by the US army's use of the defoliant Agent Orange during the war that ended in 1975.

Although I have seen a very successful plantation of an Albizia species in Sabah (North Borneo), I have also seen countless plantations that were in such bad shape they were useless — damaged by fire, drought, heat, poor planting practices, insects, or diseases. The good examples are so few and scattered in comparison to the failures, it is hard not to feel gravely disheartened by what I have witnessed. What always amazes me, however, is how proud foresters are of their plantations, which in comparison to the size of deforested areas, are pathetically small.

The Festival of Songkran, Thailand, 1983. Over the last forty to fifty years, local people in Thailand have argued that the clear-cutting of forest cover after World War II has caused a shift in monsoon rains, especially in the northeast of the country. This shift, whatever the cause, has had a major impact on natural regeneration of Thai forests.

When the spring rains fail to arrive on time here, as they do now more than three years out of every four, planting is delayed, crops suffer, and the sun becomes unbearably hot. When such conditions exist, trees shed their leaves and their seeds and hills and mountains literally disappear from view when fires fill the air with smoke. It was a surprise to me, in 1983, when rains finally came to northeast Thailand and cleared away some of the smoke. Until then I had not realized that the landscape where we lived was dotted with black limestone crags.

No wonder an annual water festival is held on April 13 to celebrate the coming of the rains; it changes everything and growth begins again. The festival is called Songkran and marks the beginning of another lunar year.

By April 13, 1983, Thailand and its people had endured nearly half a year with little or no rain. When the heavy monsoon rains did come it was August—a whole four months later than expected. This, apparently, was an unusually long time to wait for relief from drought and heat. It was normal for monsoonal rains to be one or two months late, but not four. That too, however, was changing with increasing frequency.

Waiting, Praying for Rain. The following story took place in a small village called Muak Lek, located on the western lip of the Korat Plateau and along the Friendship Highway between the cities of Saraburi and Korat. During the Stone Age, the people of northern Europe prayed for the return of the sun every spring. Now, people from dry areas all over the world pray for rains to rejuvenate their land, indicating that the ancient belief in prayer still moves us today. After my experiences of the Songkran in 1983, I realized that this festival was both a prayer and supplication to higher powers for the return of rain.

That day in 1983 started like any other—dry, hot, cloudless, and hazy from smoke. The air had a slightly brownish tinge to it and smelled acrid. As it wafted through our bedroom windows on the gentle morning breeze, it felt mildly irritating to the throat. The quiet was interrupted by a rooster's call and light began seeping into our awareness. This croak, hardly a rooster's call at all, was always followed by the call of other roosters in the neighbourhood. But none of them sounded like the first. Each morning for over two months he was the first one to wake my daughter, Hilary Louise, and me, and we quickly learned to recognize his call. It really wasn't that hard.

He sounded like he was being strangled as he crowed, ending his call in a drawn-out gasp. Maybe someone had it in for him. Perhaps he was blamed for announcing another day of toil and sweat. It was most likely, however, that he was very old and suffered from laryngitis. His call brought a smile to my daughter's face and mine as we stirred ourselves into action every time she came to visit from Bangkok, where she attended an American high school.

At the time, I was working on an ASEAN-Canada Forest Tree Seed Centre project in Muak Lek. The project was funded by the Canadian International Development Agency (CIDA). I was there as a tree seed technologist and was scheduled to stay for six months. During that time I was to work on problems relating to seed ripening, collection, storage, and germination, and to teach Thai foresters about seed technology.

My wife, Hilary, and son, Timothy, had joined me by this time. All of us ate breakfast that morning in the small cafeteria across the lane from where we lived in Muak Lek. With help from the Danes, this area was developing as a significant dairy farming area.

After breakfast we walked into the village and were immediately splashed with water from a passing pickup truck. There were about six young men on board and they were throwing water on everyone with great abandon, laughing and shouting as they did so. Small children with dishes of colourful clay came up to us, hoping to paint our faces with pink, green, and white as a sign of respect for their elders. Their faces were already painted but because they had also been doused with water, the clay was running in streams down their faces. We were a sea of coloured clay. Hair clung to our wet scalps, and our clay-stained clothes stuck to us as if we had showered fully dressed. We really looked quite exotic. The children's faces were totally unrecognizable but beneath the smears they smiled broadly.

The splashing got wilder as we approached the temple (*wat*) in the central part of the village. Soon everyone was soaked and looked a mess. There was a rule that for women, water could be splashed only on their backs, and gently, but it was soon forgotten. No one could escape the exuberance of that morning.

Throngs of people came from every direction and were moving steadily towards the temple grounds. Loud music was played by a lively band of young men dressed in wild colours. They danced around on an elevated stage. The temple courtyard was covered with people dancing, laughing, and horsing around. There must have been hundreds of people. The noise was deafening and the crowd grew steadily.

Thanks to all the water and clay, my leather sandals had become slippery, causing me to slide around in and on them. The leather thongs were stretching further and further. Within an hour they clung more to the side of my feet than stayed under them.

Then, quite suddenly, a woman grabbed my hand and hauled me into a group of older women from the village who were dancing in a ring. Half drunk with the spirit of festival and fun, my companion would not let go of my hand. Soon I lost sight of my wife and son and could only assume they had been hijacked by other dancers elsewhere.

As I danced and stumbled, my sandals were more off than on my feet. I don't think we managed to actually "dance" to the music, but we were in constant motion all the same. We laughed and laughed and the women sang while I hummed. My feet were sore but that didn't seem to matter much. It was only later that I had to tend to them when the gravel became too rough on my skin.

At first I did not recognize who these women were. They all looked like painted strangers, but soon I started to feel a sense of caring and acceptance from them that I had not felt from anyone before. It was intoxicating. Round and round we waltzed, holding on to each other so that we would not fall down laughing, all the while singing, giggling, and smiling at each other. It felt as if I was on some kind of a high, floating in a sea of caring. The people accepted us, and we were all enjoying each other's company. It was wonderful.

Soon I realized that I knew some of the women. They worked in the tree seed centre nursery where I also worked. In 1983 foreigners were still a bit of a novelty in the village, so to be dancing with a pale face was probably a special event for them.

In time we all let go of each other's hands because everyone was exhausted. We said our thank-yous, bowing our heads towards each other, palms of our hands pressed together and the tips of our fingers touching our noses. Again and again I said, "Khop Khun ma crap" (Thank you very much). Women laughed in response to my awkward pronunciation.

The festivities went on well into the afternoon, and the music blared forth without letup. After I found my wife and son, we were so tired, dirty, and uncomfortable in our wet clothes that we decided to go back to the house and clean up by early afternoon. Upon leaving we were splashed again with more water and made sure that our mouths were closed. The water was definitely not for internal use.

It has occurred to me many times since then that we were part of a much larger event. The celebration was a plea for relief from the heat and a prayer for rain so crops could be grown. I also believe that one reason we received such a warm welcome from the villagers that day was because we looked different, and they hoped that our presence would somehow add something special to their prayers. Here we were, total strangers in a small village, yet we were having such fun together. It was obvious that our willingness to join made the Thais happy. That is what it felt like.

As fun as the celebration was, there was also a serious aspect to this prayer for rain. It was not a celebration of the rains that had arrived, but for the rains to come. It was not about certainty, but about uncertainty. I felt indeed as if I had been part of an ancient ceremony, pleading with the gods—or maybe it was the giving of thanks, assuming that our prayers for rain were heard. It was that power that gripped me and made me feel so in tune with these people. I was part of their prayer but I did not understand this until much later when our forestry studies had provided some interesting results.

Tree Seed Ripening Hampered by Drought. The tree seed centre that was established in the Muak Lek village was put there to serve as a seed supply centre for the entire ASEAN region once plantations could be established. As a result I decided to study seed production and seed ripening in order to set guidelines for when and how tree seeds were to be collected, tested for viability, stored, and used.

Virtually no rain had come by February in 1983 and it soon became clear that tree-seed ripening was being affected by drought. Many tree species were starting to drop their seeds well before Songkran in April. In fact, that year they started to shed their seeds and leaves in late February. This scenario repeated itself in 1984. Some years the rains can still come by February, but in the twenty-two years between 1980 and 2002, it happened only once, according to the weather records kept at the Thai-Danish farm.

It came as a surprise to me, therefore, that seed ripening could be a problem in the tropics where it is always warm. But moisture shortages and heat worked there much like low temperatures in the fall, limiting seed ripening in temperate climates.

It was only then that I was told that large tracts of forest had been cut after World War II. Forests were also cleared for agriculture after logging was finished. The foresters I spoke to said that as these forests were being felled the spring rains started to come later and later. It was assumed there was cause and effect but there was no proof. Some people say that monsoon rains cannot be shifted by human interventions and are driven by differences in temperatures between cooler, larger oceans and the land as it heats up in summer. Any belief in the negative effect of forest clearing on climate will, therefore, be hotly debated until we know more. In Muak Lek, however, change had come and many local species were unable to produce mature seeds due to the lack of timely rains. That meant relying on human intervention to raise seedlings in nurseries, when or if good seeds could be procured.

One fact that cannot be disputed is that forests also shade the ground from the glaring sun. When trees are cut the climate near the ground gets much hotter and drier. Soil surface temperatures could exceed $80°$ C at mid day in Muak Lek, and air temperatures were an average of $10°$ C warmer at chest height in the open than they were under continuous forest cover. This makes the human environment much more uncomfortable and the environment for tree seed germination almost impossible.

By logging the forests of the northeast, Thais nearly got rid of their serious malaria problem. The World Health Organization of the United Nations had described the region as a special malaria area during the immediate post-war period. But once the forests were gone this threat to human settlement and health all but disappeared. Forests conserve moisture, and this allows mosquitoes to breed. By clearing the land, it became much drier and this prevented the mosquitoes from finding water in which to lay their eggs. The sal forests of southern Nepal were also logged to get rid of malaria. Most of the leeches in the Nepalese foothills bordering India were also lost thanks to forest removal. That, too, was a blessing to both humans and animals. Not all of the changes were positive, though.

One of the serious drawbacks of forest removal in warm climates is that it leads to loss of soil productivity. Because nutrients are leached out so quickly by heavy rains, most of the nutrients needed for good plant growth are retained in the standing vegetation. This problem leads to swidden agriculture practices. As soils get depleted, farmers have to clear more land to grow crops because returning to their original fields is not an option until they have been left fallow for many years, or until farmers can afford to buy and apply fertilizers. The result is more forest clearing and more loss of forest land to agriculture. It is a vicious cycle that leads to hotter conditions and makes it still more difficult to establish tree species.

A great preponderance of tree species in warm climates is programmed to produce seeds at the start of rainy seasons, as in northeast Thailand. Trees cannot change their behaviour to wait for late arrivals of monsoons. So if rains are delayed for whatever reason, trees start to shed both leaves and unripe seeds just to stay alive.

In February 1983, for example, one species of acacia (Acacia catechu) native to northern Thailand started responding to water stress by withholding water from one branch and then another. By Songkran, all branches were bare and all seed pods had dried and opened, leaves withered and dropped, branch by branch. Pods changed colour quickly from a deep, reddish brown to a dull brown, opening and shedding their seeds as leaves dropped. These first seeds were unripe and shrivelled up, dying almost immediately upon hitting the ground.

Gradually, these trees entered into a drought-imposed dormancy. By Songkran (mid-April) they were completely leafless and had shed all their seeds. The same happened in 1984: what little sprinkling of rain that arrived before Songkran evaporated almost as fast as it fell. Five out of six species studied for their seed ripening in 1983 at Muak Lek were found to produce unripe seeds. Seed ripening, therefore, appeared to be a common rather than sporadic problem because of the late monsoon rains.

As the drought deepened in 1983 and 1984, the seeds shed by this species of acacia showed increasing ripeness the longer they could be retained on the tree. This gradient of seed ripening during drought is an interesting phenomenon. It means that this acacia species, and others, were, in effect, hedging their bets. They were producing seeds with every possible variant of ripeness and, therefore, dormancy. This meant that they would be able to capitalize on almost any twist of climate. Unfortunately, the seeds of this acacia had been badly infested by a beetle that laid eggs in the immature seeds before the seed coats became hard. In 1983, the feeding damage caused by the larvae of this beetle had killed most of the seeds of this species.

Of the six different species we tested for seed-ripening, only one seemed able to cope well and produce what appeared to be mature seeds by Songkran. This species was a leguminous species tolerant of droughts. All other five species developed seeds with white, soft seed coats or belonged to the sal group of trees that produces seeds that must germinate upon maturation.

Species that normally grow in closed, mature forests (like the majority of the sal species) suffered most from the drought. They could not regenerate themselves naturally in the Muak Lek area because even in wetter years, when rains came early, the germinating seeds would surely die during the in-between dry years. As a result, even if the trees produced ripe seeds, they could not survive without humans to collect their seed and grow them in nurseries under shade. Their future is, therefore, very much in doubt in this area of Thailand.

The rainfall records for the village of Muak Lek show that during the twenty-two years between 1980 and 2002, heavy rains (those producing more than one hundred millimetres in a month) came in one year in February, three years in March, two years in April, twelve years in May, and one year in June. In other words, rains came after Songkran in eighty-two per cent of the years between 1980 and 2002. As a result, trees showed signs of moisture stress that was leading to seed loss.

Today in Thailand it is illegal to cut virgin forests. But for what purpose should these trees be kept except as examples of the past? Many indigenous species that cannot regenerate themselves naturally now have therefore become our serfs. When they grow old and die no new seedlings of the sal species will replace them except if put there by humans. These old, scattered trees are, for all intents and purposes, misfits in their own home. Some people call them the living dead.

Even with laws in place, illegal logging of indigenous forest cover still continues and forest lands are shrinking in the area. Karlberg (1983) wrote an article in the *Bangkok Post* in which he said that thirteen million hectares of forest land had been converted to agricultural land between 1960 and 1983. This represented fifty per cent of the country's total forest land base. He claimed that ninety-five per cent of the firewood and sixty per cent of the sawn logs were still being cut illegally by 1983, presumably from public lands. This kind of illegal logging is by no means confined to Thailand. From what I have seen it is endemic in Southeast Asia because governments are too underfunded, too corrupt, or too unaware of the magnitude of the problem to take action.

It is therefore not an accident that the tree nursery in Muak Lek in 1998 produced mostly very drought-tolerant and hardy species of leguminous trees for reforestation as well as town and city plantings (Hellum et al., 2000). Much more than fifty per cent of the seedlings belonged to this group of leguminous species.

No one wants malaria to return to northeast Thailand. No one wants the leeches or malaria back in southern Nepal, either. No one would reforest cleared land now that it is needed for food production. No one can wait long enough to grow trees as large as they once were. No one asks what our new forests, made up of aggressive species, will look like in the future or question if they will become pests that no one can get rid of.

We must also deal with the fact that some aggressive shade-intolerant species have trouble regenerating themselves under their own shade. When they grow large and are cut or die what will the ensuing regeneration look like? Will foresters be faced with planting forests from now on because natural regeneration cannot succeed? Or will all logging have to be done just as seeds mature and start to be shed? I don't think this is possible for logging that goes on all year.

Indigenous, or primeval, forests are places of mystery. They cater to our imaginations. They are endlessly diverse, silent, and beautiful. Finding old signs of human incursion in such forests, old stumps, overgrown trails, signs of old habitation and fence lines that are sinking into the earth, only make them more intriguing. These signs of failed human incursions make forests seem more powerful than we are. Given time, they can swallow up what we try to change.

Waiting and Longing for Water. Water shortage is something every human being understands. We need water every day of our lives. But as we clear more and more forest land the focus on water will sharpen as temperatures rise and the land becomes hotter and drier. River flows that are normally modulated by forests will also become more violent and sudden as forests disappear. Droughts will eventually become more frequent.

That is why I often think back to that Songkran day in 1983 when people were urging on the rains. We were dancing to attract the attention of the rain god, to tell him that rains were overdue. That is what people told me. We were united in a common prayer. But although this gave me a wonderful feeling of place and power it also signalled that things were not right with the world. It felt like participating in an ancient ritual. And why shouldn't we? The stone-age people of Scandinavia carved ships in stone to carry back the sun in spring, so why shouldn't the Thais dance to bring on the rains? The worship of natural events and causes still flourishes in the minds of many people today. I, for one, am glad.

If rains indeed become more irregular or shift entirely in time and place, humankind will face a truly difficult problem. This world has so many people that perturbations in climate or weather wreak havoc everywhere, because people live everywhere. It is a sober reminder of our smallness in the face of change. As we change the vegetation on this planet to suit our own needs (removing millions of hectares of forest cover to do so), we are also removing ourselves from our spiritual past and losing our connectedness with nature (Harrison, 1992).

That is why writers like Trefil (2004) are so misguided when they argue that we can and are managing the world's forests and lands well and sustainably. Trefil's confidence that we know enough "to garden this universe" as we please is symptomatic of our ignorance about the connectedness we are losing. Trefil thinks we can manage the earth, as he says, by people for people. Thinking this way is abysmal. All we need to do is to look around our own neighbourhoods to see how we have lost forests and diversity of environments to other uses. Diamond (2005), Fagan (2004), Flannery (1994), Glennon (2002), Harrison (1992), Quammen (1997), and Williams (2003), among many others, warn us that we have done poorly in the past and not gloriously well, as Trefil insinuates. The topic is receiving increasing worldwide attention. Some authors even question the abilities of governments to administer forest resources (Wood, 2004). Wood says that governments cater to short-term interests and not to the needs of the future, placing the future in jeopardy for our descendants.

Negros and Cebu Islands, The Philippines, 2001. The islands of Negros and Cebu lie in the middle of the Philippine island archipelago facing the Sulu Sea, about ten degrees north of the equator. Both islands have suffered loss of forests having placed major emphasis in the past on sugar production to sustain local economies.

The Spanish started to grow sugar cane in the Philippines well over one hundred years ago, growing large amounts for the export market. Some families became fabulously wealthy, having acquired the best land. Crops were harvested until the 1970s, when the country lost dependable sugar markets to the United States, which was developing its own sugar-beet industry and becoming independent of sugar imports. It was then that estate owners in the Philippines lost their incomes. In the city of San Carlos on Negros Island, both sugar mills had to close. The workers lost their jobs and moved to the forests to survive by growing their own food. Because there were few social safety nets in the Philippines, these workers had to fend for themselves. The wealthy landowners had to generate new and different crops for new and different markets. This took a long time. They had to diversify to carry on business. This process of crop change was too slow to meet the needs of the poor and jobless. The forests fell victim to land clearing by the dispossessed.

At the same time that sugar markets were dwindling, the forests of the Philippines were being felled and sold. What trees the huge and powerful logging companies (many of them American and Japanese) did not remove were subsequently cut by illegal loggers. Timber is still being logged illegally today, even in parks and nature preserves in the few remaining areas where larger trees can still be found. Local people go into the hills with their oxen and water buffalos, drag out these logs, and sell them wherever they can find a buyer. This allows them to make ends meet. These squatters, or their descendants, are still squeezing out a living on ever-diminishing resources. Some have moved off the land and become absentee landlords, renting to others less fortunate, which, of course, reduces further the incentive to improve conditions. The basic problem for the unemployed was that there was no help from society. They were left high and dry. Philippine laws state that squatters can be removed from the land they occupy, but this does not happen very often. Where else could the needy people move? In 1983 I saw one area on Luzon Island where people had been forcibly removed only to come back again and again, creating very unsafe working conditions for everyone involved.

On Negros Island cultivated areas with a slope up to sixty per cent are tilled by hand. The land is so steep in places it is hard to imagine how people could even find footholds to work the land. Heavy rains cause sheet erosions that wash the tilled soils into the rivers. In addition, the re-growth of native vegetation is repeatedly harvested for fuel-wood and charcoal, fodder, and building materials. What used to be forest land is now degraded to farmland or what ecologists call pioneer environments, where natural plant succession cannot proceed because of continual disturbances. What often saves the day for local people is that the soils of much of the Philippine islands are volcanic in origin and can stand enormous amounts of abuse while still producing good crops.

The Baticulan Watershed. Farmers in the Baticulan watershed (upstream from the city of San Carlos on Negros Island) have created a damning scenario. Their soils wash into the river that dumps its overburden in San Carlos and in the delta as the river enters the sea. Sea life is damaged this way. Catches of shellfish drop and fisheries are damaged. The citizens of San Carlos are also affected and have to cope with annual flooding and siltation. The rainy season lasts from June to August, and the typhoon season lasts from October to December. Over the past twenty years the occurrences of typhoons, and their torrential rains, have increased many times. I was told these heavy rains used to come once every few years. Now they come several times each year. They wreak havoc with soils that are steeped and tilled, and with roads, cities, and homes. An effort is being made, however, to plant trees to try to restore the Baticulan watershed. The goal is to reduce damages to the watershed, fishing grounds, and the city, and to improve the lot of the poor people who depend on the land for their sustenance.

The Imitators. To manage the forests of the Baticulan well we have to take action and do the right thing by Mother Nature. In most cases it is obvious what needs to be done. But, what is really needed is the will to improve conditions. To restore damaged environments, we need to take pride in what we do. Then we can begin to address the issue of restoration and to maintain biodiversity.

The 105 families living in the Baticulan watershed, upstream from San Carlos, have not only depleted native biodiversity and practiced farming in a way that led to subsequent flooding downstream. They have also tried to restore or replenish their depleted environments with species of plants from abroad. The watershed is small, only about four hundred hectares, and should have well over one hundred different tree species, based on what tropical forests of this kind can support. As far as I could see, however, the local people are planting only three tree species: two from Central America and one from India. They plant virtually no fruit trees except some poor custard apple trees, bananas, and a couple of jackfruit trees. They had not planted flower gardens or vegetable gardens, except to grow the corn, bananas, and coconuts that provided most of the food for their diets.

Trees taken from somewhere else surely need not replace native tree species in a species-rich country like the Philippines. Local imagination and knowledge was sorely lacking, which is not only a Philippine problem. This borrowing attitude can be found everywhere—not only among poor people but also among people who should know better than to think that their species are not good enough. It is the old problem of finding greener grass on the other side of the fence.

Pride in local trees is surprisingly hard to find no matter where you care to look—Thailand, Malaysia, China, Vietnam, or Guyana. Even in New Zealand and Australia, where they have planted North American conifers extensively, this failure to acknowledge the local vegetation is real. If we don't know what our native forests can offer us we certainly are not in tune with them. Not knowing the species and their qualities means we have lost contact with the natural world around us. Introduced species are brought in to meet various needs. Long-fibred conifers provide good, strong pulp and paper in areas where such tree are lacking (as in Australia and New Zealand). Fast-growing trees are introduced to meet urgent needs for wood as well as for domestic and industrial reasons. To introduce them is one thing, to manage them is quite another. The difficulty is that we often cannot anticipate what these introduced species will do to local flora, countryside, and public opinion.

The lack of local biodiversity in the Baticulan watershed was surely brought on by too many families moving in to farm forests in this small watershed. Biodiversity of this watershed is not improved by importing species of plants from far away. It is improved by using native and adapted species because the land is, or was, rich with these species and could become so again. Biodiversity may also be improved by introducing plants that make for a better human environment, more shade, some beauty in the form of flowering trees, more economic return from fruit trees, and crops that provide the local people with hard-earned returns for their labours. Even the corn I tasted in the Baticulan was starchy and bland. Sweet corn was apparently not available.

Change Away From Sugar Production. By 2001, some of the sugar cane lands on Negros had been converted to other crops. They were growing okra for export to Japan. They were also expanding their production of crops for sale locally. These crops included watermelons, tomatoes, onions, and pepper. This change to new crops had taken about thirty years to put in place, from 1970 to 2001.

The need for restoration of the Baticulan watershed was caused by the collapse of export market, and people slowly came to realize that almost total reliance on exports was not wise. People were beginning to cater more to local markets. But it was the forest that suffered. Foreign logging companies were invited to log the forests, and squatters moved in to survive the sugar market collapse. Once settled they remained. This situation is common wherever I have been in Asia and Guyana. The Aboriginal people who live in the indigenous forests have been displaced. Outsiders who did not have the same priorities or concerns about protecting hunting grounds or forest cover have moved in. The indigenous people do not always save the forests they call home, either—especially when they follow swidden agriculture methods. The outsiders, however, never felt concern or attachment to the land, nor were they concerned for the maintenance of the resource. They came to mine it and use it for their own survival. Gone was the feeling of respect for the forests; gone was the feeling of caring for the resource itself. Exploitation is a single-minded business.

Cebu Island. A watershed on the neighbouring island of Cebu had similar problems of erosion and siltation twenty years ago. The Mag-uugmad Foundation Inc. was established to introduce conservation ideas to protect the depleting water resources for the city of Cebu. It is still short of water, but only because the city has grown considerably in the past thirty years. A manager was hired and started to train farmers and then to train trainers to give small landowners help. Eighteen years after the project got under way in the 1970s, the farmers were much better off and led a more diverse life than before — financially, personally, and environmentally. They looked purposeful and carried themselves with a sense of pride. There was considerable farmer interest in these approaches because examples of success are irrefutable. There was also a strong sense of community here and a growing industriousness. The changed image of who these people were and what they had to offer strengthened their self-esteem. However, some farmers have yet to join and instead hang onto old ways of farming. Those ways include tilling up and down slopes, thus accelerating soil losses during heavy rains.

Farmers here grow a large variety of crops, ranging from fruit trees (such as ranbutan, which have fruit-like lichee), to pepper, coffee, orchids, and vegetables. They have been shown how to terrace their lands and build berms to hold the water and soils in place during rains. They plant a special species of Australian grass along the berms of these terraces. It grows where it's planted and does not spread. The grass can be harvested for fodder and the roots hold the berms in good shape. I was told of two brothers who had started to work with the MAG-UUGMAD foundation at its inception. One of them could now afford to send his two daughters to university. I saw chickens and pigs in this community, as well as oxen and water buffalo. The houses here were better and larger than in the Baticulan, reflecting a better resource base and a healthier people. Many species of forest trees had been planted, mainly around their cultivated fields. Reforestation per se was not possible because most of the land was now under cultivation. Nevertheless, this area was well shaded and not nearly as hot and exposed as in the Baticulan. It took eighteen years on Cebu Island to reach this stage of diversity and prosperity. It took almost thirty years on Negros Island to change from sugar to okra and other crops. In other words, it took almost one human generation to change things and find new markets for new crops, and to thus create new jobs for local workers. It also took about the same time to destroy the forests in the Baticulan watershed, to clear the land, and to end up with extreme poverty and many environmental problems. Even though the Cebu Island project was a success for human settlement and for water conservation, it was not aimed at creating natural environments but human environments. It was creating environments to serve man.

Lessons Learned. Forest degradation is occurring everywhere that people have to eke out a living with little or no support from society. Major damages can result from just twenty to thirty years of overuse, and this overuse worsens as gradual changes goes unnoticed. Regaining lost diversity takes as long again, if it is possible at all. In the meantime, conditions deteriorate. It is easy to suggest that degradation should have been prevented in the first place. That certainly would have been the least expensive alternative and would be possible if people were given help when their jobs disappeared. But population pressures are increasing throughout the world and forests will continue give way. Williams (2003) has no illusions about continued forest land losses, especially in the tropics.

Liaoning Province, P.R. China, 1999–2001

Tieling City Prefecture. Two very different examples of degradation of the environment through loss of forest land and forests come from the province of Liaoning in China. In one situation there was the virtual disappearance of a common peat moss species from the local flora due to overcultivation of flat ground. In the other example, the unfortunate practice of planting very large stands of single species of trees led to infection and death from insect damage.

In the year 2000 I went to Liaoning to advise on forest nursery practices. People wanted to learn more about how we grew tree seedlings for reforestation in Canada, a method with which I was quite familiar. One of the first things I wanted to show the Chinese was that we use peat moss as a rooting medium for growing tree seedlings in nurseries. It is cheap and readily available in Canada, and I had been told that peat mosses also grew in Northeast China. I suggested that we look for peat bogs where this moss could be found. I knew from seeing a book on local flora that one particular peat moss species (Sphagnum fuscum) was found in Liaoning. I suggested we go and look for its habitat. We spent an entire day driving around looking for suitable habitats but found none. All level ground was now farmed, tilled, and drained. Places that might have been suitable had disappeared. We did not venture into the border country up against North Korea because, for safety reasons, this area was out of bounds for most people. I don't know if the moss still survives there, but everywhere else I could find no suitable habitat. For all intents and purposes, this peat moss had been eliminated from the local flora. I had failed in my efforts to introduce this method of growing seedlings to the Chinese.

The second example from Liaoning concerns the very extensive planting of a single species of pine on vacant public lands, on the orders of the Chinese leader in the 1940s and 1950s. He had urged people to plant native trees. People collected seeds of a local pine species, called Pinus tabulaeformis, grew seedlings in nurseries, and planted them on any vacant land they could to improve the depleted forests of this province. People complied with this worthwhile venture and in the Tieling City prefecture alone over 650,000 hectares of this pine were planted. But a scale insect from Japan was accidentally introduced ten to twenty years after these plantings were completed. These pine forests today are about seventy per cent infested with this insect. This infestation leads to tree deformity and finally to death. Over 490,000 hectares of forest are in dire need of help. There is little or no money available for insect control, however, and in many places the trees are so misshapen that they are useful only for fuel. Planting such large areas with a single species invites problems. If insects or diseases gain access they can do considerable damage because the food supply is so large. Here again is an example of a failure to understand how ecosystems work. Such large homogeneous areas of forest almost never occur in nature. Some foresters have never understood this simple ecological fact. Outside inputs into local forestry practices fail so often I am tempted to suggest they should never be attempted. The Chinese now have a huge job of getting rid of this pine and replacing it with other species. Wrong decisions in forestry can have far-reaching economic consequences. In fact, I suspect that such a large forestry problem cannot be solved easily and will drag on for many decades without redress because of lack of funds and commitment.

A Lack of Harmony. Land is under pressure from competing uses throughout the world, and everywhere forests yield. When I accidentally came upon a landscape in Liaoning where I saw a balance between forestry and agriculture, it was such a relief from my tormented feelings about forestry. Here the hills were forested, the flat valley bottoms were tilled and maintained with care, and the small villages were strung out, hugging the interfaces between forest and field. It was a most beautiful, pastoral countryside. But this kind of landscape, in my experience, is rare. Normally, there is, or will be, a war between forest and field. I remember a woman once saying to a group of foresters in Alberta that she knew for sure that any forest area with good soil would have to be given to agriculture. She was proud of her influence over forestry personnel.

We rarely seem able to raise crops and maintain forests in harmony with climate and native vegetation. When faced with pest invasions, agriculture uses herbicides and insecticides. Fertilizers are employed when soils are not productive enough for our needs. Varieties of plants are used that are able to withstand extreme environments or extreme applications of herbicides, or that can produce more than native varieties can. The same is beginning to happen in forests. Things cannot be left as they are because our needs are increasing. More control over our environment is needed because the productivity of the natural environment is considered too low to meet our needs. When I come upon examples of harmony between agriculture, forestry, and society it always makes a huge impression on me. I ask myself if these examples are just fleeting visions of what might be. Alberta, fifty years ago, had a white zone (private land), a yellow zone (an agricultural expansion zone), and a green zone (for forestry and watershed protection). Now the yellow zone is all converted to agricultural production or private ownership and the green zone is being nibbled away. Decisions to log are not made locally but in board rooms far away. Industrial development focuses on creating jobs and making money, on controlling destiny and on securing investments. Protecting the environment is often considered an unnecessary or troublesome expense.

By trying to control everything in nature we cannot be in harmony with the earth. We are just beginning to understand what sustainability means in forestry because environmentalists have reminded us, again and again, that forests are more than wood factories. But I doubt that most foresters and environmentalists know what ecological sustainability implies, either in terms of biological necessity or in terms of forestry practices.

Governments need to take a more active hand in the long-term management of forests and all renewable resources because of their great value and basic usefulness to society as a whole. As you drive north from the capital of Pahang state, Kuantan, in peninsular Malaysia, into Terengganu state, you can see what happens to an indigenous forest after logging. It becomes, in large measure, converted to palm oil production. Palms now stand row upon row along endless miles of travel. As our planet gradually becomes less diverse, in species of both animals and plants, forests become, more and more, places to pillage and places to change into something else. Growing human populations will assure this destruction. And foresters are aiding and abetting this process by their silence and their complicity. Foxes are indeed in charge of our chicken coops.

Populus balsamifera
Black poplar

Pinus flexilis
"Limber pine"
Crowsnest, Alberta
September 20, 2002

In the Jungle at 3 PM, Guyana, 1994

There is a vast difference between an old-growth tropical rain forest and a forest that has come back after logging, but judging by what I have heard and seen, most people who visit the tropics never know the difference. The indigenous forests are simply too far away from tourist places, or they are too rare for people to see them. However, once you have been in such an old-growth forest you will never forget the experience. It has a grandeur and mystery, beauty and power greater than regenerated forests anywhere. It feels like being in a vaulted cathedral. Its majesty invites silence.

I want to share an experience that happened on two visits to a forest that was halfway between an old-growth forest and a second-growth forest with signs of old logging.

The events in this story took place on two separate days in Guyana, where I was working on a forestry project in 1994. I visited the Moraballie Forest Reserve for the first time on a sunny day. Our Amerindian guide led us through a beautiful young forest thinned and managed long ago by an English forester by the name of Fanshawe. With its plant diversity and richness, I found the place utterly spellbinding. Seba Creek wound its way slowly westward through the great stillness, emptying into the Demerara River some distance away. In places it flowed over white sand and in places over red laterite clays. There, in shafts of light, the waters looked as red as if this was a river of blood. The colour came from the extraordinary amount of tannic acid from decaying plant materials. These red waters tasted harsh but were safe to drink. Amerindians said: "Drink of them and you will never want to leave."

Over the slowly flowing waters grew fern, dipping their fronds into the water, waving in the currents. Flowers hung from vines drooping in the still air. Colours of red, yellow, and purple mingled with the deep green colours of leaves. It was beautiful yet almost sinister because of the colour of the water. It was idyllic, very still, graceful, and forever changing. The sun passed overhead, casting shadows that moved like wands over the ground. I came back to Georgetown looking forward to my next visit.

Several weeks later I visited the same area again with about a dozen men and women. We had come to see if this forest had escaped logging in the past, and if it was worth saving as a research and protected forest area. We also wanted to learn what Amerindians had managed to learn about forest inventory procedures (cruising) from our project staff.

The Moraballie Forest Reserve had been set aside as a research forest many years ago by Fanshawe, but was said to have been logged by poachers who apparently had moved the boundary markers.

To get to the Moraballie Forest Reserve we travelled south from Georgetown to Linden. Then we travelled from Linden further south, leaving behind the infertile "white sands" soils along the coast and entering a region with a mixture of white sands and red, clay-rich laterite soils that are typical of tropical regions worldwide. The sun was shining. It was morning and the air was still cool. We were all looking forward to our hike in the fragrant air of the forest.

The Reserve now had mostly second-growth trees as a result of illegal logging and the fact that Fanshawe had started experimental trials in the area. The trees were tall, healthy, and of good form. It had been selectively logged for Greenheart (Chlorocardium rodeii) for probably close to fifty years. The branches of the canopy above cast almost complete shade, except along some parts of our trail. We walked in the coolness, avoiding the bright, sunlit spots where it was hot and humid.

I have been told that the docksides in New York harbour are built with Greenheart poles because this wood is resistant to teredo insects that commonly bore holes in and destroy poles and pilings standing in brackish water. The wood of Greenheart was, therefore, of great commercial value. The forest that was left behind after logging now looked healthy. It was no longer composed of the "most popular Greenheart," and this may explain why the area had been left alone in recent years.

Some of us were dressed in bush clothes and boots for this hike, but many wore city clothing and shiny shoes, combs in pockets. The ground was dry and the day was sparkling bright without a cloud. We walked along an old so-called snake trail (an old road left behind by the loggers), what we in the West would call a skid trail. We also walked along blazed trails where the northern boundary had been marked on the ground. Our guide was an Amerindian of the Arawak tribe.

On our walk we saw many blue butterflies with wingspans of about five inches (12.5 centimetres). The wings were iridescent blue on their upper sides and black underneath. When they flapped them we saw flashes of blue light one second and then nothing the next, as their wings opened and closed above their heads. The butterflies tended to congregate in spots in the forest where the sunlight reached the ground, along the snake trails where we walked. There were dozens of them in the shafts of light. When they flitted about, flashing their blue wings, it felt like a psychedelic experience. If we had been hearing loud music it would have made the experience complete. But the forest was silent except for the occasional squawking of the hornbills that flew by with their heads down. These psychedelic moments were beautiful and surreal.

Around two in the afternoon we finally reached the end of our hike, sat down on old logs, and started to eat our lunches as we chatted. We had come to a spot maybe two kilometres, as the crow flies, from where we left our vehicles at road's end. We talked about the changes that the illegal logging had brought about but decided that the area was still worth keeping as a research forest, mainly because of Fanshawe's work years earlier.

Suddenly it started to cloud over and drops of rain fell into the dense foliage overhead and descended to the ground as mist. We had hardly finished eating when the moving clouds darkened the sky in patches. The light started to fade as heavier clouds moved overhead from the east, invading our cathedral-like world. We were in the belt of the Trade Winds that blow so steadily at this latitude.

We decided to head back to the vehicles as fast as we could. But we were far from them. These clouds looked like they meant business, growing darker by the minute. Our Amerindian guide suggested we should make a beeline for the vehicles, cutting cross-country to do so. We asked if he could find the way and he nodded and smiled. He started walking east through the bush, and we followed single file behind him.

We walked through uncharted bush without any trails. We strode faster and faster as our guide began to jog ahead of us. The rain came down in torrents and darkness fell upon us as we jogged through open and dense patches and under lianas drooping akimbo from taller, leaning trees.

The sky grew even darker and rain fell in increasing torrents. The air was alive with hissing as water hit foliage and bent it. As leaves rebounded they were hit by more rain. The soil under foot turned soft and muddy after many of us had trodden the same ground. Wood became slippery, which was a concern considering that crossing the Seba Creek on moss-covered logs would normally scare me. But during this jog I didn't think about it as we ran across them and darted under lianas and leaning trees, trying to avoid the grasp of thorny bush and rattan. It didn't occur to us then but later I thought about how I could run over these logs in the rain and darkness when, under other circumstances, I would fear falling into the water. How could my balance be affected by my mental state? But it was.

By three in the afternoon it had become so dark I could hardly see ten feet ahead. We had to jog fast to keep up with our guide, yet we could hardly see where to plant our feet. I started to pass people because I didn't relish the thought of losing sight of him. And he didn't look behind to see if we were following. He just jogged along. I passed only a few people before I had to give up. Besides, we were all moving fast enough that passing wasn't really possible anymore. The bush was too dense. The darkness was too limiting. I just prayed we wouldn't lose sight of our guide.

But no one got left behind. No one got hurt or was so out of condition that such a jog did them in. Our world had become a soupy darkness full of hissing and thrashing noises as if we were inside a massive orchestra playing tunes on the leaves and branches. Composers of atonal music probably would have been delighted by this racket. We didn't give a thought to the danger of snakes, poisonous centipedes, spiders, leopards, panthers, or wild boar. We just jogged. I remember tucking my field notes into the little plastic-covered pocket in my Tilley hat because that was the only place left that was still relatively dry. As it turned out, the notes were wet anyway by the end of our jog, but they were at least legible enough to be transcribed later.

We were amazed when we finally came out of the bush about one hundred feet from where the vehicles were parked. Then the rains suddenly stopped, the sun came out, and it turned hot and sultry. We looked a mess, all of us. We wrung out our clothes and didn't seem to care who was watching as we stripped. But more than being happy that we had made it out of the rain and darkness, I was impressed with our guide's ability to come out in the right spot. How had he done it? After all, the country was fairly flat, there were no clear landmarks anywhere that I could see, and the darkness had precluded any of us judging direction by where the sun was in the sky. Only the Seba Creek flowed in the general area, making many twists and turns, and could hardly help direct our journey. I went over to our guide to thank him and asked how he had known how to come out where he did. He just looked at me as if to say it was nothing at all. Why was I surprised? He lived close to nature and forest.

I would never have believed it if I hadn't experienced it, but at 30° C, still wearing my wet clothes, I could hear my teeth chattering. I was freezing cold. For the entire trip back to Georgetown, a drive of about two hours or more, I sat hunched over the car heater trying to control my shaking and to keep my teeth from rattling in my mouth. "You should always bring along spare clothing," Nadeira said as she fetched a dry T-shirt from her bag which had been left in one vehicle.

I would not have believed, either, that such darkness could occur in mid-afternoon had I not experienced it myself. When the rains stopped and the sun came out, I began to wonder if I had imagined everything. We had moved from sunshine and light, to almost total darkness, and then back again in what must have been no more than an hour's time. Could plants survive under such conditions during prolonged monsoon seasons? How did they maintain life in near darkness? The on-again off-again photosynthesis that must take place here, even under sunny conditions, is astounding. Sunlight reached the forest floor only in fleeting sweeps as the sun moved overhead. That anything could survive here during these dark times is a marvel. During the monsoon seasons the darkness can last for days and weeks.

Because the Greenheart tree is one species that can grow here, it is worth explaining how it seems to do it. Its large and heavy seeds (about the size of flat golf balls) fall to the ground and can lie there for months without germinating, which is possible since they are so poisonous that nothing eats them. These seeds can germinate in total darkness, and the seedlings can grow in almost complete darkness for up to two years, relying almost solely on the stored food in the seed. By the end of the two-year time the seedling can be six feet tall, with hardly a lateral branch and only a few dark, small, green leaves at the very top of a pole-like shoot. The whole stem and leaves can photosynthesize because all parts are dark green, right down to the ground. This is the ultimate climatic climax species, to my understanding, meaning the very end of plant succession from a pioneer to a mature and self-perpetuating condition. Growth can start in deep darkness and every time the forestry canopy opens up, as old trees fall down or branches break off, Greenheart grows quickly in height to wait again for another gap to come along after the canopy above closes again. If this tree finally reaches the canopy high above it starts to produce large, long branches that shade out all competition for great distances. I have been told that such a single mature tree can subdue competition and dominate areas up to an acre in size. But for any single tree this growth may take several hundred years. No one really knows how fast these trees grow and how old they can get because tropical trees do not put on annual growth rings that can be counted. When these trees grow old and start to decay, their seeds have fallen below them and new trees can grow as they themselves did long ago. As a result, they could become permanent residents, dominating such sites, if larger disturbances did not come along to really open the forest to sunlight. If the forest is opened then Greenheart loses out to other species and forest succession starts all over again. Greenheart does not grow well in open areas. The trees grow well only once they reach the top of the forest canopy.

The hapless people who came to the forest in their city clothes and shoes looked bedraggled, to put it mildly. Some tried to clean their shoes which were covered in red mud, leaves, and twigs.

Some tried to comb their hair after wringing out their clothes. Even though I had worn my leather boots and bush clothes, so that all I had to do was wring out my clothes and empty my boots of water, I didn't look much better than the people who dressed for the city. I shook with cold too, something no one else seemed to do for they had brought spare clothing. My boots have never been the same since.

I felt immeasurably fortunate to have experienced two opposite situations. The first was a day of special beauty and serenity and then, in the same general area on the very next visit, the day was filled with darkness and torrential rains. Was there any better way to demonstrate the need to be in the forest to experience its different faces? I held the forest in greater esteem than I had before, and my respect was tinged with fear. I was glad not to have had to spend a night there under those conditions. Extremes show the dimensions of any problem, and the more I learn about these forests, the more I am in awe of them. I have of course experienced torrential tropical rains under other circumstances, but no place has filled me with more respect and awe than the Moraballie Forest Reserve did.

Another time, on Negros Island in the Philippines, and in the torrential rains of a nearby typhoon in 2001, I experienced what sheet erosion is like. It means that so much water is falling that all the ground is covered in many centimetres of water at one time. That rain came down so heavily that a creek we had just crossed dry-footed quickly turned into a torrent a metre deep and full of brush, leaves, and mud. It was hard to cross, even when holding onto lianas and using poles to steady ourselves, but we had to get back to our vehicle. Waters came to my waist as I staggered across. Rain flowed down from the surrounding gentle hills in sheets ten centimetres deep or more. Thank goodness the soil here had not been tilled. Maybe then we would never have been able to cross this small creek. The soil stayed in place fairly well, but the creek waters, nevertheless, were laden with silt and clay.

Such situations have to be experienced. No amount of reading can create the kind of respect and understanding that field encounters can. Thus we get to know our forests, learn to respect them, and grow in our skills to manage and look after them.

Populus tremuloides
Aspen

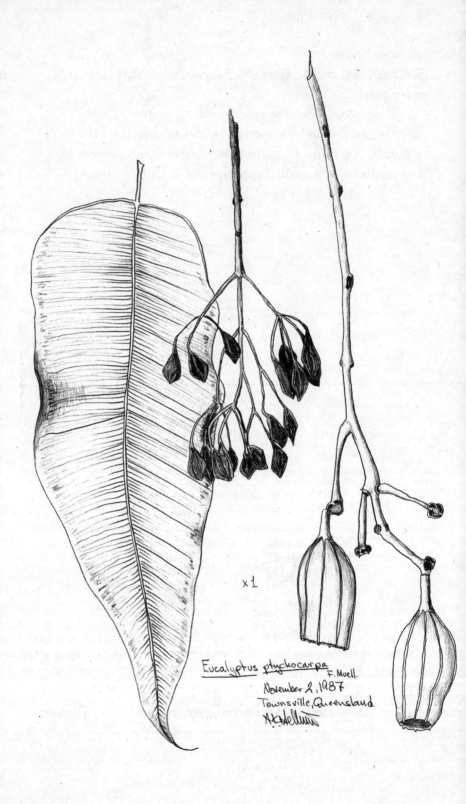

Eucalyptus ptychocarpa F. Muell.

November 2, 1987
Townsville, Queensland

x1

Wealth in the Form of Diversity

There is a lot of discussion these days about preserving biodiversity, but it is very difficult to define what this means. People try to define the term (Kimmins, 1997), but the more they try, the more convoluted and impractical the definition becomes. Foresters think of trees, botanists think of plants in general, zoologists of animals, and tourists of vistas, exotic wild animals, and undisturbed environments. The range of special interests is endless.

Biodiversity means too many different things to too many people. This is why we need a broader definition. I think the word diversity would be easier to understand. We relate to this word in our daily lives. The more diversity we have the richer our lives become. More diversity in life means more options and more opportunity. We can do more, travel more, and learn more from the world. We can think more creatively and, hopefully, become more compassionate with and more tolerant of others. Because the word diversity is less abstract than the word biodiversity, maybe we could see more clearly how important it is to preserve nature in its diverse forms. Perhaps then we could understand how important it is to preserve the web of life in any ecosystem.

The problem with that plan, however, is that we want to grow single crops so as to control the environment and to minimize the presence of weeds or pests. We know intrinsically that to do so is unnatural because any natural ecosystem is complex and made up of variable organisms, each one pursuing its own survival and propagation. So we create more and more artificial environments in agriculture and in forestry, and spend increasing amounts of money to maintain them. We know also, even as we pour chemicals onto our fields to maintain crop monocultures, that diversity leads to stability in our lives and in nature.

The single crop, the homogeneous countryside, fields and fields of cane, canola, wheat, and extensive forests of single tree species are inherently vulnerable to pest attacks, diseases, market downturns, and overproduction for what markets can absorb. It is easy to understand that diversity in such situations creates more stability for both people and environments. The people of Liaoning in China also discovered that planting so much pine of a single species exposes the species to a grave threat of infestation. Had the Chinese been able to plant many different species, rather than just one, the scale insect would probably have had far less impact on the local countryside and on the local economy, even if all the pines of this one species had died.

Forest ecosystems in north- and south-temperate zones are simple compared to tropical forest ecosystems. Part of the relative simplicity of these temperate ecosystems is due to the many ice ages the earth has experienced over the past millions of years. Species were forced to move or died as ice and cold advanced. Temperate species got squeezed into more tropical climates and found themselves stranded as climates again warmed up. Not enough time has passed since the last glacial retreat 10,000 years ago to develop in temperate forests the same complexity that exists in tropical climates that escaped glaciation. Species migration, as ice retreated, is often slow and incomplete, even today. For example, spruce trees are still not native to the west coast of Norway. They will grow there if planted, but they are not native. They haven't had time since the glacial retreat to migrate across the mountains. Migration can take thousands of years. The Norway spruces that were brought from other places and planted there cannot produce seed. They are adapted to grow where planted but the climate is unsuitable for seed production. For re-vegetation to be successful under natural conditions, adaptation and seed production must happen in step with migration.

The diversity, or biodiversity, in our forests in Canada is still so complex that foresters have trouble understanding how to manage them ecologically and sustainably. The fact that we think mostly about trees in forest management, and little else, is a sad reflection on our state of awareness and caring for forest ecosystems. We are too stuck in the exploitative mode. We do not understand how complicated the interdependence is between organisms, the interdependence that creates stability and sustainability. The species of trees we plant in Western Canada today, and especially in British Columbia, are only modest versions of what grew there when we first logged. Next time we log such forests we will have much smaller trees because we cannot wait for them to grow large. Even this north-temperate "simplicity" is too complex for us to understand. We are unaware of what a forest ecosystem really entails because we are so singleminded in our pursuit of profit from trees. We could do much better by listening to trees, getting to know what they need, and what they depend on before we cut them down. Then we wouldn't be left wondering why they regenerate themselves poorly. Our Canadian forests go through a succession of restorative phases after any disturbance — not just the regeneration effort we want to impose.

The first colonizing plants that enter after a disturbance are called pioneer species. They prefer open habitats with little competition. They grow best in full sunlight and, normally, grow very quickly. In Alberta we have willows, aspens, poplars, and several species of pine that prefer these disturbed situations. The spruces grow under the shade of colonizing plants, and after long periods of time the true firs get established. Such a sequence might take several hundred years in Alberta. Foresters call these last species climax or climatic-climax species because they can regenerate themselves in their own shade, something that pioneer species have trouble doing. About half of our forest cover in Alberta is aspen and poplars and about half is pine and spruce. The true firs comprise a minor component of no more than maybe twenty per cent of the spruce forests. The reason for this distribution is that Alberta experiences, and has experienced in the past, regular wildfires that destroy hundreds of thousands of hectares of forest annually. Fire sets back the succession of species much like logging does.

When we then try to manage forests for those species that have greatest economic value we interfere with the natural plant succession that any forest undergoes in its quest for maturity. We have interfered this way for centuries in Europe without major problems because we did things on a much smaller scale than we do today. Now we do things so fast and on such a large scale that attention to detail is impossible. Because we don't pay enough attention to details, we don't adequately match species to site and climate. The bigger our units of operation, the greater our problem will be with adaptation to place and climate. The closer we get to the equator, the greater the problem becomes because in tropical forests this succession of plants, from pioneers to climatic-climax species, is opportunistic (Oldeman, 1990) and rarely predictable. That is, it is haphazard, depending on the availability of seed at critical times and on the right environment.

For years, foresters tried to regenerate the Greenheart tree in Guyana after logging. But they failed because they did not understand that this species does not like open sunlight or open growing conditions. More than anything it needs intermittent light and darkness to grow tall and straight. This species, therefore, does not lend itself to management by humans. Large Greenheart trees take a century or more to reach maturity, and that does not fit the model for rapid tree growth that we have set as a standard. I feel a great respect for the tenacity and competitive strategy of this species. But if it does not lend itself to "management" by people, will it survive our logging? I am glad to say that listening to this tree, to figure out its strategy for survival and growth, has given me great joy. Here is a prime example of a tree that defies human intervention. We stand in awe of it and struggle to master it. But our chance of success in small because Greenheart does things in its own way, and this way does not fit human time-frames at all. Greenheart is greater and more complex than our management skills can handle, and we must respect that or we could eradicate it altogether. The species is vulnerable because its behaviour does not fit our timeframe for logging. Great care is needed to manage this species well. Species extinction is a real danger if we fail to consider its needs.

Schumacher (1973) wrote convincingly about why Small is Beautiful because he understood that we need to pay attention to diversity in nature by reducing the scale of our operations in both agriculture and forestry. He argued that we need to think smaller to get off this spiral of exerting ever-increasing control over natural processes. We need to change our whole mindset about how we view nature and what it can give us.

For one thing, we have to curb our inveterate desire to copy others—to make every place similar to every other place—by planting crops in our backyards because they grow well in the yards of our neighbours. The picture is complex and I am not sure that I have a clear message to give except that we do have to listen more to what Mother Earth and her trees, other plants, and crops tell us about themselves. Only when we start to think ecologically will we be able to talk realistically about ecological sustainability. That is the middle way. We cannot, in the long run, use force to strong-arm nature into meeting our needs.

Loss of Diversity at Home. Diversity is not dwindling only in the trop-ics. It's also dwindling right here in Alberta, where I live. Because I am a painter of plants I now have to drive north into the boreal forest or west into the mountains to find the plant diversity I look for. In other words, I have to travel beyond the farming fringes. Just thirty years ago I could find many different plant species to paint right near Edmonton, but not anymore, with some notable exceptions where special environments have been preserved or left alone. The natural diversity of the prairies has been reduced through land cultivation and through our diligent use of fertilizers, herbicides, and insecticides. By mowing grass along road ditches and right-of-ways, filling in sloughs, and clearing native brush for housing developments and crops, we have further reduced natural diversity. Natural habitats are being lost everywhere. We may have brought about less change to our environments than in many other parts of the world because we have been farming here for a shorter time than elsewhere, but we are on the same road. We, too, are losing much of the plant and animal diversity as we strive to control nature.

I planned to cut down several patches of Canada thistle in bloom on my land the other day. But when I got there, these patches were being visited by dozens of butterflies taking nectar from purple blooms. Should I leave the thistles for the butterflies and entertain the ire of my neighbours and county over my spreading weeds? Or should I wait and cut the thistles when the flowers no longer yield honey? I decided to wait a few days. Compromise can also be honourable.

Think of wealth as not being defined by money, power, or control but by the mind's perception of options and possibilities—by our understanding of this world and its limitations, as well as opportunities. The earth's resources are not limitless, even if our exploitative imaginations may be. This is our home place, to use Stan Rowe's title. We don't exploit our homes if we plan to stay around for long. But we are blind when it comes to the exploitation of public lands. This is a beautiful world. It can offer us, our descendants, and our imaginations immense possibilities through mindful, sustainable use. We cannot keep on losing fifteen million hectares of forest land annually forever and hope to preserve diversity or ecological stability. Nor will we be able to enjoy the forest's magnificence.

Abies balsamea

Balsam fir

Astelberria

Vaccinium myrtilloides
"Blueberry"

Terratima, AB.
June 19. 2002

The Strangler Fig

Forests play the role of our mute partners in life. We rely on them for building materials, fuel, oxygen, recreation and hunting, medicinal plants, and plants of great beauty.

They define the boundaries between the known and the unknown. Workers find employment in them and forests protect the soil from erosion, provide organic matter, and help control river flows. All these forest functions provide reasons to protect them. They are essential to us.

But forests have also been seen as our enemies, standing in the way of agricultural and urban expansion and harbouring wild and dangerous animals and people. Is this the reason we find it so hard to preserve and protect them? Many have little empathy for them.

Even as we need forests more and more, they disappear in favour of almost every other use imaginable. Humans are experiencing nearly exponential population growth today. That is why in Southeast Asia alone, in the 1980s, over two million hectares of forest land were lost annually. FAO (1986) reported that we lost 11.3 million hectares of forest land annually to other uses worldwide. Williams (2003), with newer and more up-to-date data, suggests this might be closer to fifteen million hectares. The question is how long this loss can be sustained without causing serious worldwide problems of wood supply. We are surely influencing oxygen and carbon dioxide imbalances, erosion, and floods, and we jeopardize the production of food for increasing numbers of people. It might look like we are heading for the same fate as the Easter Islanders or the Norse on Greenland centuries ago. We are facing climate changes today which may or may not be entirely of our own making; we are losing significant parts of earth's forest land to other uses, mainly agricultural, and we are repeating patterns of behaviour demonstrated in recent history when societies foundered. Because I live in a consumer society, I cannot help being as much a part of this problem as anyone else. But I try to reduce my consumption of resources as much as is physically possible. It makes me angry and despondent, as well as frustrated and painfully aware of how dependent we all are on services. We rely on investments that earn us money, cars that take us where we want to go, services that pollute the environment, building materials that house us, foods that have come from far away, non-renewable resources that heat our homes, the newspapers that keep us informed, and the TV that brings us stories from all over the world. I feel as if I am in a spider's web.

This steadfast pressure on forests everywhere reminds me of the strangler fig in tropical forests. The strangler fig uses its host tree to grow large and then it strangles it. Birds eat figs, carrying them up into the tree crowns to eat and leave behind seeds that germinate and start to grow epiphytes—plants growing on plants. The long aerial roots of strangler figs grow down along the branches and trunks of their hosts or hang down like curtains until they reach the ground. There they enter the soil and start absorbing and transporting nutrients and water. As these fig vines grow larger they start to strangle the host, which then dies and rots away. By this time the figs can stand on their own because the aerial roots have grown large and have coalesced into what looks like a solid tree trunk. They then look like trees in their own right.

Strangler figs can grow to great size, with trunks over six metres in diametre. In the middle, where the host used to be, they are hollow. Some hollows are so large that people can climb inside them, right up into their crowns, depending on how large the original host was. These fig "trees" are still found in logged forests. They are there because they have no value to the logger. Their wood is useless for lumber.

It has irritated me when tour guides on several occasions have extolled strangler figs as remnants of the original forests. They are remnants, to be sure, but they do not represent the beauty or grandeur of the original forest. A mature, closed tropical forest looks and feels like a cathedral. The strangler fig "tree" has no grandeur. Its crown is a mess of smaller, thinner branches. There is sparse growth on the forest floor under old tropical forests because the light is so dim that few plants can grow there. The tree crowns cover the heavens way above your head, and the trunks of such trees can be fluted or round, and their branches are usually immense and gnarled. Strangler fig "trees" have relatively low crowns compared to regular trees and do not at all give a sense of space and majesty. They can look quite grotesque.

The strangler fig uses its host for its own purpose. When the fig starts to become independent of the host, the host dies from strangulation. Humans, like strangler figs, use the trees and forests for their own purpose, without concern for the host itself. To the fig, the host has utilitarian value in the same way the way the forests have utilitarian value to humans. But while the strangler fig develops into an independent "tree," the forest industry will collapse without access to wood. In the same way the strangler fig loses, or kills, its host, we are losing our forests because of the way we log. We are also losing the very wood we need. In Canada we have already logged somewhere around forty per cent of our forests, according to one report. The loss is both physical and spiritual, even if many of these logged forests are now regenerating themselves with and without human help.

By making forests more and more dependent on us for their own survival, by changing environments for regeneration, and by growing trees like crops, we shape them more and more to fulfill our own needs, while knowing too little about theirs. We can control forests and manipulate them to our own ends. We don't look up to them any longer. They serve us and that is all. Trees are now considered beautiful mostly because of the wood they produce, not because of their form, colour, or age. We lose sight of the beauty of creation in its whimsy, its force, and its richness.

I have given examples of forest mismanagement to demonstrate our failure to see forests as living ecosystems in their own right. Surely, if forest management is to serve public needs, we need to change the way we see forests and how we use them. Were we to consider forests as discrete ecosystems we would argue against current standard forest management. It needs a fundamental shift in our thinking. It says that forest growth has to be regulated, managed to produce as many possible specific crops as possible in specific timeframes, all to serve man in every way. If forests are to survive as our partners, this attitude, first formulated in Germany centuries ago (Harrison, 1992), has to change.

Our inability to identify tropical tree seedlings explains, to some degree, why we plant exotic species so often, rather than relying on native regeneration, which is seen as too slow and imperfect. We think we can find quick answers in forestry by planting "super" trees. People think they can reap quick profits by growing trees more quickly than before. Potentially fast-growing trees planted in the wrong places will fail because we overestimate what is possible. Our hubris is alive. Then we lose interest and try other species, leaving the failed planted trees as scars on the countryside. Mistakes are repeated again and again because of ignorance and because of bravado and cocksureness. I sometimes think we should stop using the slogan of "a seedling planted for every tree cut," which gives a completely inadequate impression of what really happens when a forest is cut. It can take a century or more to get the new stand back up to logging standards even if regeneration has been successful. The world has seen many successful examples of reforestation (Evans, 1985), but we rarely hear about the failures. To see them you have to get off the beaten track and into remote areas where visitors rarely venture. I have seen too many plantation failures to feel confident that we know enough or care enough to manage forests sustainably, even where land uses are not diverted into profitmaking.

What we take from the forest land base we do not give back easily. When I was told in 1986 that Agent Orange killed almost all native forest cover on over 650,000 hectares in the Mekong Delta of Vietnam, I found it hard to believe. The land was sterilized this way, my translator said, to expose the Vietcong forces during the Vietnam War. Once the vegetation was killed, it burned; and now that sedges and rushes have invaded the land, it is doubtful if it will ever again have significant forest cover. Too many other interests have taken over the land for that to happen, and major efforts at drainage are now needed. Some drainage had already started in 1987 but reeds cover the land evenly and densely. The job of reforestation will therefore be difficult and expensive, especially when the Vietnamese, in the years after the end of the wars, could not get seed to do the job because of an international embargo. Not only were the site-adapted species lost in this massive destruction, but after the war Vietnam could not obtain seeds from abroad to start reforestation. Their own seed-source stands had been lost. It was not until 1986 that FAO of the United Nations became involved in helping, eleven years after the withdrawal of American troops. By that time the whole vegetation complex had changed from forest to reed swamps.

When I went to the Mekong Delta to see what could be done, the Vietnamese took me by canoe to see a reforestation project. I stood on the gunnels of this canoe and could barely see over the reeds to verify that, in fact, forests were pretty well wiped out as far as I could see. It was an upsetting experience to look around and see almost complete deforestation in a 360-degree circle around me. Ditching was going on to dig large canals for draining the floodwaters and to raise the adjacent land above flood-water levels. People had moved in and were trying to cultivate some of the higher ground, but any effort to reforest such an expanse boggled my mind. The floodwaters of the Mekong River are acid so the soils of the delta are also acid, and the water is not potable. Drinkable water was transported in by canoe for people to buy, ladle by ladle. This is what happens in the name of war. It is so dispiriting because forests and local people are always the losers.

The Mekong Delta area that I saw was transformed by the use of Agent Orange. The trees were gone, the reeds had taken over, and the small efforts to reestablish trees gave no sense of promise. The only species of tree that people knew how to grow here was the paperbark tree (Melaleuca cajuputi). Its seeds can germinate under the acidic flood waters and can grow upwards until the seedlings poke above the receding water. Then they change form altogether. Going to a reforestation project in a canoe was a very strange experience, and learning about the regeneration capacities of the paperbark tree certainly stretched my understanding. This world is so full of wonders—the paperbark tree certainly fits my understanding of a miracle tree. Maybe there is still hope for reforesting such vast areas destroyed by man if enough seeds can be procured to do the job. I left feeling complete uncertainty about what the future might hold for this area.

One strategy is to forget what we had and cope with the present as best we can. The Vietnamese told me to forget the war and to focus on the future. This moved me to silence for my mind was anywhere but in a forgiving mode.

At the end of a professional career, I feel eternally thankful for the gifts that forests have afforded me directly and indirectly. I feel privileged to have been awarded the opportunity to relate to forests. My advice to future foresters is to rekindle that intimate feeling—it can be called love—that is needed for us to care for forests in our trust. We desperately need to be professionals rather than just employees. That implies defending sound, ecological forestry principles in the face of economic pressures. It is important for our own sake, for the sake of our children, and for the sake of the forests themselves. When we distance ourselves from forests we don't know them well enough to manage them sustainably. With time and with growing distance between forests and ourselves, we are losing our respect for them, for the power and good they can provide. Because trees are pressured into becoming our servants, our lack of knowledge of them, and our lack of compassion for them, will become a threat to their and to our own existence.

Robert Pogue Harrison (1992) said in his book *Forests: The Shadow of Civilization* something akin to my saying that humans are hollow inside like the strangler figs. I quote from his wonderfully sensitive and insightful book:

> The global problem of deforestation provokes unlikely reactions of concern these days among city dwellers, not only because of the enormity of the scale but also because in the depths of cultural memory forests remain the correlate of human transcendence.

He goes on to say:

> We call it the loss of nature, or the loss of wildlife habitat, or the loss of biodiversity, but underlying the ecological concern is perhaps a much deeper appreciation about the disappearance of boundaries, without which the human abode loses its grounding ... Without such outside domains there is no inside in which to dwell.

In other words, since we are bound to keep on logging our forests it is important to regenerate them well and ecologically and not to lose our sense of boundary between the forests as places of wonder and our urban worlds. If we lose the sense of boundary that forests give us, a boundary between the known and the unknown, between the safe and the dangerous, between the variable natural environment and the confines of cities, we lose our sense of centre, our sense of who we are. This is what I mean by saying we are hollow at the centre of our beings when we destroy forests the way we do. We become more and more like strangler figs when we fail to maintain natural, stable boundaries in our lives. It is as true for the child who needs to know boundaries to develop his or her centre as it is for people to know that wild places define who and what we are.

A drungtso (doctor of traditional medicine) in Bhutan told me in 1989 that he had studied medicine in Tibet for seven years. During this time he had been taught how to understand nature before trying to treat its ailments. He said:

> You learned that not only do plant species vary in their ability to heal human frailty, but you also learned that each species, of the many hundred species we studied, varied in their potency by location of where they grew. Soil type and its richness, the moisture and its seasonal patterns, the altitude and its aspect or exposure, as well as temperature, were factors influencing potency. Medical potency always increased with altitude for any given species. In addition, to practice Tibetan medicine you had to learn how to compound the ingredients from many different species to achieve a good healing result. Then you could dispose of all books for then you knew your trade.

I laughed at the thought of throwing away all my books, having accumulated many during a life of study. The thought seemed frivolous to me. When I had a chance to read Sennett's book *The*

Culture of the New Capitalism (2006), it dawned on me that here was a man who said the same thing about current thinking in business and government. Sennett points out that today it is not what you know that has value but how innovative and flexible you are. This approach to problem solving and creativity, he argues, can lead only to shallow thinking because to dig deeply into any subject is today interpreted as being stuck and rigid in one's thinking. The culture of new capitalism discounts experience and historical perspective, stressing the need to be innovative. This short-term way of thinking, he argues, leads to emotional stress and excessive worry about the future. It leads to insecurity and worry because boundaries and guidelines are missing.

Upon reading this passage, I immediately thought of the Amerindian who led us through the jungle in near darkness without losing his way because he was so well rooted in his world. He performed a miracle in finding the way. And I thought of the time in China when I was given less than twelve hours to prepare myself before presenting a three-hour lecture on tree seeds. Everything I was to say had to come right out of my head. How could I possibly do my best under such circumstances? It seems to me that, for my own survival, I would take the experienced, well-rooted approach of the Amerindian over my impromptu lecturing approach any day.

Two views, the Tibetan and that of New Capitalism, share the belief that they can dispense with all books for, at some final stage, we know enough to make good judgements at the spur of the moment and can, therefore, be creative in our work. This creativity sounds like a throwback to medieval thinking to me, for we are — and always will be — rooted in our past. To ignore that is to court the possibility of social revolution or deep psychosis. In our headlong gallop into the future it is hard to listen. But without listening to trees, how can we see them as our partners in creating a better world?

Salix sp.
willow

References

Braathe, P. 1953. Investigations concerning the development of Norway spruce vegetation which is irregularly spaced and of varying density. Medd. Norske Skogforsk. 12: 209–301.

———. 1976. Investigations concerning the development of Norway spruce regeneration with irregularly spaced and varying density. 2. The stability of the zero-square percentage. Medd. Norsk Inst. Skogforsk. 32: 505–520.

Brody, H. 2001. The other side of Eden. Toronto, ON: Douglas and McIntyre.

Bugeja, M. 2005. Interpersonal divide: The search for community in a technological age. New York, NY: Oxford University Press.

Cattaeno, A. 2002. Balancing agricultural development and deforestation in the Brazilian Amazon. Washington, DC: International Food Policy Research Inst., research report no. 129 (abstract only.)

Darwin, C. 1868. The domestication of plants. London, UK: John Murray.

Diamond, J. 2005. Collapse: How societies choose to fail or succeed. New York, NY: Viking Press.

Evans, J. 1985. Plantations in the tropics. Oxford, UK: Oxford University Press.

Fagan, B. 2004. The long summer: How climate changed civilizations. New York, NY: Basic Books.

FAO. 1986. Tropical forestry action plan: Committee on forest development in the tropics. Rome, Italy.

Flannery, T. 1994. The future eaters. New York, NY: G. Braziller.

Glennon, R. 2002. Water follies: Groundwater pumping and the fate of America's fresh waters. Washington, DC: Island Press.

Harrison, R.P. 1992. Forests: The shadow of civilization. Chicago, IL: Chicago University Press.

Hellum, A.K. 1994. A seedling identification guide: Trees of Guyana. Edmonton, AB: Lone Pine Publishing.

———, P. Pukittayacame, M. Kashio, and S. Kijkar. 2000. A field guide to some tree seedlings in Thailand. Bangkok, Thailand: Craftsman Press.

———— and F. Sullivan. 1990. Symbiosis between insects and the seeds of Sesbania grandoflora Desv. near Muak Lek, Thailand. *Embryon* 3(1): 37–39.

————, Zainudin M. Ariff, and S. Ibrahim. 2005. *A field guide to some tree seedlings, medicinal plants and mangrove species in Pahang and Terengganu states, Malaysia.* Pahang, Malaysia: Pertcetakan Utama.

Homer-Dixon, T. 2001. *The ingenuity gap: Can we solve the problems of our future?* Toronto, ON: Vintage Canada.

Karlberg, D. 1983. The denuding of forests: Why we need a new approach. *Bangkok Post,* 27 February 1983, p. 7.

Kimmins, J.P. 1997. Biodiversity and its relationship to ecosystem health and integrity. *Forestry Chronicle* 73(2): 229–32.

Klamkamsorn, B., and T. Charuppat. 1983. Forest situation of Thailand in the past 21 years (during 1961–82). Bangkok, Thailand: Remote Sensing and Forest Mapping Subdivision, Forest Management Division (in Thai).

Langor, D.W., and J.R. Spence. 2006. Arthropods as ecological indicators of sustainability in Canadian forests. *Forestry Chronicle* 82(3): 344–350.

Langor, D.W., J.R. Spence, H.E. J. Hammond, J. Jacobs, and T.P. Cobb. 2006. Maintaining saproxylic insects in Canada's extensively managed boreal forests: A review. In proceedings, insect biodiversity and dead wood. Asheville, NC: Dept. of Agriculture, Forest Service, Southern Research Station. Tech. Rep. SRS–93, pp. 83–97.

Mabberley, D.J. 1987. *The plant book.* Cambridge, UK: Cambridge University Press.

Oldeman, R.A.A. 1990. *Forests: Elements of sylvology.* Germany: Springer-Verlag.

Quammen, D. 1997. *The song of the dodo: Island biogeography in an age of extinctions.* New York, NY: Simon and Schuster.

Schmidt, R.C. 1984. Seed production and tree improvement in Puerto Rico. Bangkok, Thailand: IUFRO, Project Group P2.01.00, May 22–26, 1984.

Schumacher, E.F. 1973. *Small is beautiful.* London, UK: Abacus.

Sennett, R. 2006. *The culture of the new capitalism.* New Haven, CT: Yale University Press.

Trefil, J. 2004. *Human nature: A blueprint for managing the earth — by people, for people.* New York, NY: Times Books.

Tshering, S., and A.K. Hellum. 1990. *Identification of some tree seedlings in Bhutan.* Dept. of Forestry, Royal Govt. of Bhutan, FAO/UNDP/UN, FAO Field Document #4. Bangkok, Thailand: Craftsman Press.

Wasuwanich, P. 1984. Seed problems in Thailand. Bangkok, Thailand: IUFRO Group P2.01.00, May 22–24, 1984.

Williams, M. 2003. *Deforesting the earth: From prehistory to global crisis.* Chicago, IL: University of Chicago Press.

Wood, P. 2004. Intergenerational justice and curtailment of the discretionary powers of governments. *Environmental Ethics* 26: 411–28.

a

b

c

Index

P

q

R

Q

t

y

u

v

w

Acknowledgements

To me, writing is a process of exploration — a search for the things I believe in and value. It is not a self-evident process, and often I am embarrassed upon realizing I may have lived a lie or been ignorant of the obvious. In any such exploration, one needs mirrors, people off whom to bounce ideas and ways of expression. Many people helped me grow during my search, and I am deeply thankful to all of them.

Had it not been for Leslie Dawson's critical comments regarding discreetness of thought, for Alan Aldred's and Alf Farenholtz's comments on content, and Gus Loman's comments on analysis, I would still be trying to find my way out of the forest. And had it not been for the kind editorial assistance of University of Saskatchewan Professor Don Kerr and copyeditor Carmen Hrynchuk, I would still be mute. But maybe more than anything, I am most grateful to the breadth of thinking that books have given me. They not only confirmed my belief in the great urgency for better stewardship of nature, but also deepened my appreciation for this opportunity to write about what I have seen.

Dr. Andreas Kåre Hellum studied forestry at the University of British Columbia and completed a PhD at the University of Michigan. He went on to teach silviculture at the University of Alberta, where he is now a Professor Emeritus in the Department of Renewable Resources. A father of three, he currently lives in Sherwood Park, AB, and works as a translator of Norwegian, his native language.

Hellum is one of few Westerners to have visited and worked in Bhutan, which is located between India and China. His previous title, A Painter's Year in the Forests of Bhutan is the result of his research and original artwork from Bhutan, which portray several species of flora rarely seen by North Americans.

NeWest Press strives to protect our Home Place, with full knowledge of the toll that publishing can take on the environment. We are always seeking ways to reduce our carbon footprint or, as the noted Canadian ecologist Stan Rowe so eloquently put it, to "Guard and Maintain the Health and Beauty of the World."

The text paper used for printing all of our latest books is acid-free, 100% old growth forest-free (100% post-consumer recycled), processed chlorine-free, and printed with vegetable based, low VOC inks. These inks are transported in reusable drums, saving a tremendous amount of landfill space.

The typefaces used in this book are FF Seria and FF Seria Sans, which were designed by Martin Majoor in 2000.

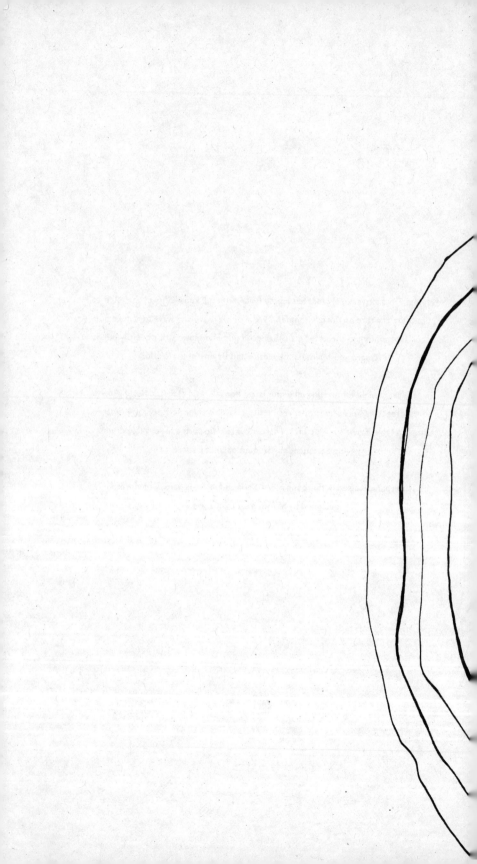